Communicating the Environment

T0316895

Environmental Education, Communication and Sustainability

edited by Walter Leal Filho

Vol. 7

PETER LANG

Frankfurt am Main · Berlin · Bern · Bruxelles · New York · Oxford · Wien

Manfred Oepen
Winfried Hamacher
(eds.)

Communicating the Environment

Environmental Communication for Sustainable Development

PETER LANG
Europäischer Verlag der Wissenschaften

Die Deutsche Bibliothek - CIP-Einheitsaufnahme

Communicating the environment : environmental communication
for sustainable development / Manfred Oepen ; Winfried
Hamacher (eds.). - Frankfurt am Main ; Berlin ; Bern ; Bruxelles ;
New York ; Oxford ; Wien : Lang, 2000
    (Environmental education, communication and
    sustainability ; Vol. 7)
    ISBN 3-631-36815-1

            Manfred Oepen and Winfried Hamacher (eds.)
                by order of GTZ - Deutsche Gesellschaft
                für Technische Zusammenarbeit GmbH:
        Pilot Project Institutional Development in Environment.

Deutsche Gesellschaft für
Technische Zusammenarbeit (GTZ) GmbH

        Division 44 - Environmental Management, Water,
                    Energy, Transport
    Pilot Project Institutional Development in Environment (PVI)
                Dag-Hammarskjöld-Weg 1-5
                    D-65760 Eschborn

                Type-setting and Layout:
        Thomas Krüger, ACT - Appropriate Communication
                in Development, Reutlingen

                    ISSN 1434-3819
                    ISBN 3-631-36815-1
                    US-ISBN 0-8204-4802-8

                    © Peter Lang GmbH
        Europäischer Verlag der Wissenschaften
                Frankfurt am Main 2000
                    All rights reserved.

                Printed in Germany 1 2   4 5 6 7

Preface

This reader is envisioned as a resource for policy-makers and project planners, providing an overview of Environmental Communication as a management tool for initiatives geared towards environmental sustainability. The authors hope that their articles will convincingly show why and how Environmental Communication should become an integral component of policies and projects, and thereby help ensure that adequate human and financial resources are allocated to this end.

The idea for this book was born at an international workshop on "Communicating the Environment", organized by the Deutsche Gesellschaft für Technische Zusammenarbeit (GTZ) GmbH in Bonn, Germany in late 1996. The participants included the editors of this reader and a number of the contributing authors. During 1997 and 1998, GTZ successfully followed up an initiative by the Commission on Education and Communication (CEC) of IUCN, the World Conservation Union in Gland, Switzerland, namely to convince the Organization for Economic Cooperation and Development (OECD) that a "Practical Orientation" on Environmental Communication should be adopted. In the course of several rounds of peer reviews by specialists associated with GTZ, IUCN, the UN Food and Agriculture Organization (FAO) and the World Bank's Economic Development Institute, the concept and contents of this reader took shape as well.

The editors would like to thank Anna Puyol of IUCN-CEC, Gillian Martin Mehers (formerly with the International Academy for the Environment – IAE), and Ronny Adhikarya of the World Bank and Winfried Hamacher of GTZ-PVI for the substantial competence, conceptual depth and critical ideas they provided during the early stages of compiling this book, and for their encouragement and active support during the later complex and crisis-ridden phases.

Last but not least, it is thanks to the generous financial support of the Deutsche Gesellschaft für Technische Zusammenarbeit (GTZ) GmbH and its Pilot Program on Institutional Development in the Environment, (PVI), in Bonn, Germany that this reader was made possible.

Wedemark and Bonn, Spring 2000
Manfred Oepen and Winfried Hamacher

Contents

PART 1 - Executive Summary

Objectives

This reader is written for middle-management planners of NGOs, government agencies and development organizations who run environmental projects or programs and know in general that communication and non-formal education are important. What they may not yet fully be aware of is the way Environmental Communication (EnvCom) can be integrated into planning and implementation. Furthermore, environmental project staff of such organizations who are supposed to put communication and non-formal education into practice may benefit from this book as it provides "best practice" examples.

The reader's major objective is to demonstrate that environmental programs can be managed more effectively if communication is systematically and strategically planned. Concepts, success stories and lessons learned relevant to policies and programs for sustainable development are "showcased". This should enhance the visibility, legitimacy and identity of those organizations and individuals already engaged in EnvCom.

Structure

PART 1 - Executive Summary provides an overview of the objectives, structure and contents of the reader.

The guiding question of **PART 2 – Orientation** is "Why this Reader?". **Hamacher, Juedes and Oepen** propagate that "society is communication and that is why an environmental problem that is not talked about does not exist" by shedding some light on the "Sustainable Development" debate, a complex topic but not an "impossible image".

In **PART 3 - Conceptual Framework of Environmental Communication**, one of the editors, **Manfred Oepen** puts Environmental Communication in a context by offering some definitions, linking it to closely related fields and cross-referencing it to other sections of this reader. Both governmental and community-based approaches are discussed in a non-exclusive manner. Because of the role of social learning, public participation and practice change in sustainable development, special attention is given to a community-based approach to EnvCom.

11

Van Woerkoem, Hesselink, Gomis and Goldstein discuss the evolving role of communication as a policy tool, from a product of government to "government as communication". The use of communication is not directly oriented to a fixed policy product, but more generally to participatory social problem-solving.

As conflicts often play a prominent role in EnvCom, **Winfried Hamacher and Karola Block** elaborate how conflict management can be employed as an alternative policy instrument that offers ways to build consensus and convergence in situations of conflictive decision-making processes. A central premise of related procedures is that by separating the negotiation process from the disputed contents or issues, communication between the actors is improved and a commonly accepted solution may be found.

Mainstreaming the environment through Environmental Education, Training and Communication – EETC is the main concern of **Ronny Adhikarya**, exemplified by the process, strategies and results of incorporating participatory environmental education and communication activities into agricultural training programs in six countries in Asia. Concrete results from these two-year development cooperation activities are demonstrated.

Oepen elaborates in **PART 4 - Environmental Communication Strategy** that the systematic use of communication is essential not only in project implementation but also to the improvement of policies and programs designed to promote participation in support of sustainable development. Communication is a two-way process in which a combination of "top-down" and "bottom-up" flows of information and experiences is required to analyze a given situation, determine the characteristics of strategic groups or the key problem to be tackled in order to arrive at the best mix of policy instruments. This is best achieved by means of the systematic "1o Steps of a Successful Environmental Communication Strategy", a step-by-step planning process in four stages (assessment, planning, production, action & reflection) from situation, audience and KAP analyses via media selection and message design to pretesting, field implementation and M&E.

The findings from eight projects or programs described in **PART 5 - Case Studies** refer the reader to a number of specific examples of how features outlined in the conceptual framework are put into practice. **Rogers et al** submit a new communication approach, entertainment-education, by means of a radio soap opera in **India**. The EETC concept developed in PART 4 is highlighted by **Rikhana et al** through a training program for agricultural extension workers in **Indonesia** designed to promote sustainable rural development practices. **Oepen** presents the case of long-term public awareness initiatives that benefited waste pickers and recycling efforts in **Indonesia**, and were built on the 10 Step Env-Com strategy in PART 4. A wide variety of environmental education and communication measures in **Africa** that focus on traditional and group media for children and teenagers is described by **Trudel**. The systematic and well-researched use of multi-facetted campaigns in support of sustainable regional development in **Ecuador** is outlined by **Encalada**. An endangered species, the black lion tamarins, in a national park in **Brazil** is being protected through a series of EnvCom activities, depicted by Padua. The model of Applied Behavioral Change – ABC as presented in PART 4 proves its validity in a Natural Park in the **United States**, advanced by **Smith**. A conflict-ridden environmental dispute between farmers and the government in the **Netherlands**, as explained by **Hesselink et al**, was balanced when distributive negotiation ("getting a share of the cake") was replaced by integrative negotiation ("creating the cake")

The contributions in **PART 6 - Lessons Learned** try to draw conclusions and lessons from the case studies in PART 5 by comparing and taking into consideration the conceptual framework of EnvCom in PART 3. They are grouped under six headings - involving the people, addressing local needs, managing a sustainable initiative, sharing responsibility through cooperation, mainstreaming the environment, and a checklist for environmental communication in projects

In **PART 7 – Resources**, a fresh outlook at work in progress regarding web-based EnvCom training, links, games and exercises is offered by **Zschocke** which may be very helpful in supporting environmental initiatives. **Martin** links key concepts that many EnvCom efforts both support and rely on: team-building, systems thinking and sustainable development. Selected **EnvCom training** opportunities are listed as well as **job aids** for each of the 1o steps of the EnvCom strategy. An extensive international **bibliography** on EnvCom literature and case studies is included .

Major Findings

Ever since Chapter 36 of Agenda 21 declared that "education, raising of public awareness and training are linked to virtually all areas in Agenda 21" (Report of the United Nations Conference on Environment and Development, Rio de Janeiro, 3-14 June 1992) many environmental and development agencies have undertaken numerous efforts to put environmental education and communication to work. Yet environmental projects and action plans often have limited

14

success because the innovations and solutions they have to offer are not fully "owned" by the people concerned. It is widely observed that environmental programs could be more effective, sustainable and significant if Environmental Communication were regularly employed. Reasons for this limited success may include basic constraints resulting from the way people think or behave:

- Assumptions on the part of environmentalists believing that scientific facts and ecological concerns are convincing and compelling on their own. However, what affected people perceive is influenced by emotions and socialization as well as by reason and knowledge.
- Inflated expectations that the "cognitive power" of the word and the image alone will solve a given problem. By taking a shortcut from "Said" to "Done!", communication barriers are often disregarded .
- Conflicts of interest which are fought by stakeholders, not negotiated by "shareholders". Confrontational approaches lead to one-way information dissemination disregarding understanding, instead of relying on two-way communication towards "shared meaning" and "win-win" situations.

Also, practical limitations related to the absence of a communication strategy lead to shortcomings. For example:

- a systematic and holistic communication strategy that takes into account people's perceptions and also saves funds is rarely considered - but it could determine the success or failure of a project,
- communication activities are often done on an ad-hoc and sporadic basis, mainly using top-down mass media while neglecting public participation in community media,
- many decision-makers do not know how to incorporate a communication strategy in their environmental project life cycles and, hence, are not willing to invest in this.

A rich body of evidence, lessons learned and best practices suggests that such limitations can be overcome by the strategic use of Environmental Communication (EnvCom):

■ EnvCom is the planned and strategic use of communication processes and media products to support effective policy-making, public participation and project implementation geared towards environmental sustainability. Embedded in a well-defined communication strategy, EnvCom makes efficient use of methods, instruments and techniques well established in development communication, adult education, social marketing, agricultural extension, community health and other fields.

■ EnvCom is a management tool, like the chain on a bicycle. The bike won't move without it but the transmission cannot move on its own. Similarly, Env-Com transforms the power generated by project managers and people concerned into action. It is the missing link between the subject matter of environmental issues and the related socio-political processes of policy-making and public participation. EnvCom is intricately related to education and training activities, bridging "hard" technical know-how and "soft" action-oriented behavior change.

■ Communication will play a crucial role throughout the policy and program life cycle of recognizing - gaining control over - solving - and maintaining control over an environmental problem. The essential aspect for a policy-maker or planner is to realize that different actors are involved at each stage, and that each actor has different perceptions, interests and "hidden agendas". Understanding where the project is in its progression from identification, formulation, implementation and management is an essential basis for determining which communication instruments should be used.

■ Many planners tend to think that producing posters and video films or launching a mass media "campaign" is a solution to problems rooted in environmentally non-sustainable practices. However, such isolated products only have a chance of success if they are integrated into a comprehensive communication strategy which defines up front for what purpose and for whom information is meant, and how beneficiaries are supposed to pop pop them into communication and action. This is best achieved by means of the systematic "1o Steps of a Successful Environmental Communication Strategy", a step-by-step planning process in four stages

Stage 1   Assessment
Stage 2   Planning
Stage 3   Production
Stage 4   Action & Reflection.

■ Empirical evidence from many projects around the world and lessons learned from the field indicate that environmental practitioners should

- define EnvCom as an output (supporting the goal of a project, e.g. "Information on Environmental Impact Assessment law disseminated") or an activity (supporting the output of a project, e.g. "Communication strategy on recycling developed with relevant actors"),
- plan the communication strategy ahead, taking research, continuous monitoring and evaluation, process documentation and an exit strategy seriously right from the beginning in project planning,
- start locally and modestly, and link issues raised, problems addressed and solutions proposed to existing trends, services and potentials, if possible by "piggy-backing",
- make use of the upstream compatibility of media, e.g. from theater to video and from there to TV,
- diversify the operational levels of a communication strategy, e.g. local theater, city newspaper, and national TV,
- use participatory approaches in media production, management, training etc. to increase local ownership and credibility and, hence, program effectiveness, significance and sustainability.

PART 2 - Orientation

Sustainable Development - An Impossible Image?

"Sustainable Development - A Communication Problem?" was the theme of a recent issue of one of the leading environmental magazines in Germany (Politische Ökologie 63/64, 2000). "Isn't that more a problem of implementation and action, not of talking about it?" was a common reaction. What many critics do not take seriously is that people and organizations rarely take action in matters they do not understand, do not consider feasible or do not benefit from. In particular, people refrain from acting if they are afraid others may not join in and cooperate, so that the "early adopters" of the innovation concerned could turn out to be the losers.

And this is exactly where communication comes in: "Without communication on sustainability no sustainable society" (Lass/Reusswig 2000:11). Or, as many sociologists put it: Society is communication and that is why an environmental problem that is not talked about does not exist (Luhmann 1986). This statement is driven by an additional dynamic in at least three dimensions of modern society:

- information and communication technologies (computer, Internet, interactive media) plus the related individual capabilities (flexibility, complex and multimedia thinking, filtering of "noise") play a more and more important role,
- politics by command-and-control decisions are outlived by an expansion of relevant actors and forms of interaction,
- society increasingly breaks down into differentiated sociocultural groups and lifestyles pursuing varying economic goals, cultural values, social codes and forms of communication and, last but not least, their own resource utilization or view of the environment.

In Germany, a country with a rather high environmental awareness by international comparison (de Haan/Kuckartz 1996: 63-82), only 15% had heard of the term "sustainability" in 1998 (Preisendörfer 1999). Obviously, sustainable development has not yet hit everyday communication. It is still a topic of deliberation among experts only. But this is not where this discussion belongs. Sustainable development, as defined in Agenda 21, means no less than restructuring modern society in a global context, in all its inter-related facets of life and by all its interwoven fabrics of actors in an environmentally sound and sustained manner. Yet, if this very society has not yet heard of the rationale of this "wind of change" - there is indeed a communication problem.

This is, in a nutshell, why communication is such an indispensable element in the debate on sustainable development. And why this Reader was compiled , starting from the notion that many would agree that, yes, communication *is* important while few would explain exactly *how* it is to be used effectively. Therefore, its target audience is primarily middle-management planners of NGOs, government agencies, and development organizations who run environmental projects or programs. They usually know that communication and non-formal education are important. What they do not yet fully understand is the way Environmental Communication (EnvCom) can be integrated into planning and implementation. Also, project staff of implementing agencies are addressed whose task is to put communication and non-formal education into practice. What they need most is "best practice" examples and references related to a systematic, step-by-step and strategic approach. The objective of the book is to "show-case" EnvCom concepts, success stories and lessons learned relevant to policies and programs for sustainable development. This should support recognition, support and replication of lessons learned in EnvCom with policy and decision-makers, and enhance the visibility, legitimacy and identity of those engaged in this new field.

Before going into the details of the concept of Environmental Communication, Winfried Hamacher, Ulrich Juedes and Manfred Oepen, for orientation, will shed some light on how this Reader fits into the context of the international discussion on "Sustainable Development", why the latter is not optimally supported by communication, and who is working hard to change this.

Sustainable Development as a Guiding Principle

Winfried Hamacher

Sustainable development as a social image or guiding principle for the future should strike a balance between improving people's economic and social standards of living on the one hand, and conserving the natural resource base on the other. This principle enjoys broad consensus both in terms of expert opinion and in common sense. Nevertheless, a closer look reveals considerable potential for conflict that clouds the viability of this consensus at the implementation level. The pursuit of sustainable development as a guiding principle has a dramatic impact on life and work, on the political and educational system. In this light, implementation appears to be an unfeasible goal, at least for the next few decades. Is the concept of sustainable development feasible by a majority? Its tenets must be more attractive to the individual than "conducting business as usual". Hence, attitudes must be changed, and the concept will have to be perceived more as a platform for debate on a future "culture of sustainability" as Juedes calls it in the next section of PART 2. In essence, sustainable development is an imperative in terms of environmental ethics, and its concrete translation into environmental objectives, priorities and options needs to be negotiated by means of discourses in various specific situations. Accordingly, there is no blueprint or prescription for sustainability. It should evolve from communication processes of consensus-building and decision-making. Sustainability will only emerge as an image shaped and determined by a given society if this communication process is participatory in nature. Hence, consensus-building and decision-making should incorporate structures and processes of social communication, political mechanisms and other elements relevant to the people's everyday life in addition to ecological and economic aspects.

When applying these requirements to the third pillar of the sustainable development paradigm, "ecological sustainability", the somewhat narrow concept of Environmental Education - which is chiefly geared to formal school education - should be broadened to encompass Environmental Communication. Environmental Communication is understood here as the planned and strategic use of communication processes and media products to support effective policy-making, public participation and project implementation geared towards ecological sustainability.

Environmental Communication thus assumes major sociopolitical functions, in particular at the implementation level. These functions are concerned with the transfer of knowledge, balancing of interests, reflection, analysis and action orientation:

- The **transfer of knowledge function** is born of the necessity to start by laying foundations of factual knowledge about the physical, chemical and biological interdependencies of natural systems and their reaction to human interventions on a local, regional and global scale.
- The **balancing function** of Environmental Communication derives from the fact that the moral premises and values, political goals and strategies for action of sustainable development harbor potential dilemmas. The right course of action is the outcome neither of the lowest common denominator, nor of an "either/or" decision. It should result from a process of balancing and reconciling diverging interests. Environmental Communication should lend support to this process by presenting a wide range of perspectives on environmental issues, and providing a forum for discussion and reflection on conflicting alternatives.
- A complex image such as sustainable development is firmly rooted in a cultural and temporal context. It does not emanate from a vacuum but from specific premises, traditions and values; it sets priorities and analyses relationships, necessitating reflection. This **function of reflection** assumed by Environmental Communication ensures that sustainable development finds its expression and concrete form in a manner commensurate with the specific context. For instance, it is essential to clarify what will be acceptable to the various individuals and groups, given their differing needs, capabilities, potential and opportunities.
- Transfer of knowledge, balancing and reflection should be empirically based on the cognitive, moral and behavioral competence that people require in order to act in an environmentally responsible fashion. Even if balancing and reflection have produced satisfactory results, the road to actual implementation is long and by no means guaranteed. In its **analytical function**, Environmental Communication can make a substantial contribution to gaining systematic information on the process of knowledge generation, on the debate on environmental issues, and ultimately on the conditions for environmentally sound practices.
- Ultimately, thinking and talking will not make a difference if communication does not lead to action. Therefore, the **action orientation function** of Environmental Communication, in a "said - heard - understood - approved - done" sequence (see PART 3), is often the most visible function as it serves as a management tool in many environmental project and programs.

Communication measures that make a serious effort to meet these functions can be invaluable in pursuing the image of sustainable development.

Goal Functions as Human Orientors of Sustainable Development

Ulrich Jüdes

The idea of "Sustainable Development", as agreed upon by more than 170 countries at the UNCED 1992, is a reaction to rising global problems in different fields, which have been discussed separately before (e.g. Brown 1993). Agenda 21 was a new step to aggregate the results of these discussions into one political document and to come to an agreement on more or less pragmatical suggestions for global, regional and local actions. This offers a vision for the future and calls for a new paradigm for both science and society. Criticism of the sustainable development paradigm (e.g. Dovers and Handmer 1993) is based on the very heterogeneous, partially contradictory assumptions and calls for a transdisciplinary comparative approach to analyze different disciplinary perspectives and aggregate their models (Jüdes 1998). An analytical approach based on systems theory is much in need that would interpret sustainable development as a cultural process and by means of a goal function concept. A comprehensive framework of this type is urgently required to prevent the further degradation of sustainable development through misconstrued use in science, politics, and by special interest groups sometimes referred to as "sustainabilism" (Gligo 1995). Moreover, the framework may help to identify suitable starting points for communication between the disciplines and for mediation between conflicting interests.

Basic Questions

Communicating the idea of "Sustainable Development" in society raises four questions of central importance:

- Which assumptions underlie the idea of sustainable development (analysis)?
- What is the theoretical reasoning behind sustainable development (concept)?
- What are the relevant fields for planning sustainable development strategies (strategy)?
- Which factors need to be considered in translating sustainable development strategies into action (practice)?

Some of the basic assumptions underlying the idea of "Sustainable Development" can be analyzed as follows:

"Development" is widely understood as a process of gradual qualitative change of a system (progressing from one stage to another), very often including an increase of complexity, differentiation and organization. But it is a cluster-term, which is used in very different contexts (biology, social sciences, politics, technology etc.). The diversity of its use and its dynamic character are responsible for different meanings which resist all efforts to find a single definition fitting all situations. A second difficulty arises from the fact that development is not related to a static condition but to a goal. Therefore it is a normative term dependent on individual and collective standards of values and varying with time and locality (Nohlen and Nuscheler 1993: 56). The RIO Report of the Club of Rome (1977) formulated six principles of human development: equality, freedom, democracy and participation, solidarity, cultural diversity, healthy environment.

"Sustainability" is derived from the Latin verb "sustinere" and describes relations (states or processes) that can be maintained for a very long time or indefinitely (Jüdes 1996). This is a general and very unspecific definition. If applied to complex situations that are based on conflicting structures or competing trends, it definitely leads to misunderstanding. This is the reason why, in the context of this Reader, sustainability is defined as a criterion for evaluation of human relations with nature and of human practices related to the environment. Human practices have influenced non-human nature to a great extent. At the same time, human relations with nature have determined the former's particular cultures and modified their environments. Thus, the (co-evolutionary) functionality of human culture for the struggle to balance human health and ecosystems health can be described as an indicator of sustainability (Di Giulio and Monosson 1996).

The use of "sustainability" as a criterion in the evaluation of the relationship between humans and nature is based on five assumptions:

- The integrity and evolutionary capability of ecosystems are basic conditions for their performance (e.g. carrying capacity).
- The diversity and performance of natural systems are environmental prerequisites for human life, and especially for a "good life".
- The utilization of nature for human living and well-being, for cultural and economic purposes, results in stress, change and damage to ecosystems.
- The performance of ecosystems can only be preserved if humans accept limits set by natural space, resources and the regenerative capability of those systems.

- The destruction by humans of structural diversity and of functions in or of ecosystems not only reduces the ecological capacity but results in new uncertainties, in other limitations to human ("good") life and in a loss of vitality of humankind on the whole.

However, a combination of fuzzy terms like "development" and "sustainability" cannot create a precise term. Hence, the meaning of "Sustainable Development" is even fuzzier and extremely resistant to clarification. More than 60 definitions are to be found in literature (Marien 1996). The definition cited most often is that of the Brundtland-Commission (Hauf 1987) which characterized sustainable development as an adjustment of three kinds of relationships:

1. relations between human needs and nature's capacity (problem of retinity),
2. relations between the needs of the poor and the rich (problem of intra-generational equity) and
3. relations between needs of the present and those of the future generations (problem of inter-generational equity).

All three relationships have been unsustainable to date, and that is why "SD" has become such a prominent political term. IUCN/UNEP/WWF (1991) describe sustainable development as a process of "improving the quality of human life within the carrying capacity of supporting ecosystems". This definition includes three basic key concepts:

- a co-evolutionary concept of humans and nature,
- the sociocultural concept of human needs and
- the natural science concept of the (limited) ecosystems.

Philosophically, sustainable development can be regarded a regulative idea in the sense of Kant. This ethical imperative tends to balance relations between current and future human needs and the capacities of nature systems. It is safe to add that this balance is not a static one. As we have learned from history and ethnology, the quality of human interaction with other humans or with nature is very specific to the given context and culture. Nevertheless, there are five principles, characteristic to men and women in all cultures: they

- react to problems when problem pressure is high
  (principle of reflexive fitness),

- creatively allow for unexpected solutions (principle of creativity),
- react to positive guiding principles rather than
  catastrophical scenarios (principle of hope),
- are able to change behavior if it is helpful to
  a group of people (principle of responsibility),
- communicate and cooperate if possible (principle of sociability).

These characteristics have functional qualities and constitute a long tradition in educational theory and practice. They can be regarded human goal functions. Human cultures are the result of these functions within specific environments. Glaeser (1992) understands environmental problems as the result of cultural crisis. Following this interpretation we may conclude that it must be possible to overcome this crisis and solve today's problems through cultural change. Therefore, sustainable development is interpreted here as a process of cultural development leading to a new paradigm, the "Culture of Sustainability".

Culture of Sustainability
The new paradigm is characterized by the intention and endeavor toward a successful contribution to the co-evolutionary functionality of human culture in finding a balance between human health and ecosystems health. Co-evolutionary functionality is defined here as the degree to which a balance between human health and ecosystems health exists. A prerequisite for measuring the co-evolutionary functionality is the identification of elements that promote the goal of sustainability. Such goal functions are defined as characteristics of all kind (elements, structures, principles) which have a promoting functional quality in a process for which a goal has been identified and accepted or constituted (e.g. human development or sustainable development).

The wide variety of ecological, economic and social conditions worldwide demand different strategies of sustainable development. Hence, contrary to the globalization trend, cultural heterogeneity in modeling sustainable development is a "Must", and the "Culture of Sustainability" requires local adaptation. From a systems perspective, sustainable development can be seen as a macroprocess of the global system consisting of an unlimited number of microprocesses of subsystems. Both scales differ in their dynamics: The macroprocess is per definition directed from conditions of unsustainability toward those of sustainability. The main processes fit given specific environmental situations and can include highly dynamic and even catastrophic events when seen on their specific microscale. This means that micro processes may be unsustainable themselves but their results may contribute to sustainability on a higher level.

This observation, well-known in ecosystem research (Jax et al 1996), so far has been widely neglected in the sustainable development debate.

An Analytical Approach to Sustainable Development

Along the lines of the questions posed above, the analytical approach to sustainable development will be structured here on four levels (Jüdes 1996):

1) a problem analysis and ethics level,
2) an epistemological-conceptual level,
3) an organizational theoretical or strategic planning level,
4) a practical level.

Fields of disciplines relevant to sustainable development can be assigned specific roles for specific levels. Studies in ecology, economics , business and industry, agriculture, forestry, science and technology, energy, transportation, tourism, architecture and town planning, health, politics and planning, social science, education, philosophy and ethics will contribute to the process of sustainable development. Yet, even though sectoral approaches are helpful for detailed aspects, they cannot deliver a paradigmatic concept of "Sustainable Development" (see Jüdes 1996). For example, each of these aspects grounds on disciplinary key concepts with specific characters (elements, structure, principles) helpful in understanding subsystem components of microprocesses of sustainable development (e.g. ecosystem, energy flow and productivity in ecology, resource use, allocation and growth in economy, human rights, basic needs and justice in social sciences). Sectoral concepts and characters are necessary for evaluating sustainable development activities according to accepted scientific standards but a sustainable development paradigm requires a holistic and bridging "common sense" that can integrate divergent ideas and models. Therefore, it should be both communicable to individuals and social groups, and operational for pluralistic reasoning and transdisciplinary evaluation of sustainable development processes.

Social Functions, Orientors and Indicators of Sustainable Development

Aside from the theoretical construction of a conceptual framework of a sustainable development paradigm, as outlined above, communication on sustainable development in society entrusts science with two more tasks: developing methods for implementing and monitoring the new paradigm. Basically, three steps lead from theory to practice:

Firstly, it is necessary to select characters (elements, structures , principles) promoting sustainable development. Characters in sectoral key concepts are of different types: characters promoting Sustainable Development (goal functions), characters inhibiting Sustainable Development (goal-inhibiting functions) and characters with functions not related to Sustainable Development (goal-differentiating functions). The list of potential goal functions presented in Tab. 1 exemplifies sectoral terms which are functional for the new paradigm on the epistemological-conceptual level. Such lists should be drawn up for all relevant sectors and problem fields.

Secondly, it is important to identify goal functions associated with more than one sector or field as they may aggregate different aspects of sustainable development (integrative goal functions). This calls for interdisciplinary cooperation. Integrative goal functions would help to bridge sectoral models and could be used as orientors in inter-disciplinary communication of a sustainable development theory. They are starting points for defining standards of this paradigm and for developing indicators.

Thirdly, among these characters another differentiation has to be made: some have cumulative and others have emergent characteristics. In the context of sustainable development, studies have been focused on cumulative characters, either with promoting or inhibiting functions (e.g. environmental pollution, toxicological substances, economic growth). These characters have been predominantly used for modeling so far while emergent characters were given less attention. This is a serious failure as emergent characters contribute to new qualities and modify the "sense" of a system, subsystem or component (Krohn and Küppers 1992). It seems to be the most critical and difficult point in the Sustainable Development discourse.

List of Potential Goal Functions Relevant to Sustainable Development

Ecological	Cultural	Social	Economic
• complexity	• perception	• pluralism	• benefits
• retinity	• rationality/reasoning	• participation	• economic production
• biodiversity	• sense/meaning	• communication	• market
• stability	• self-formation	• social integration	• innovation
• energy storage	• religion/myth/sanctification	• pro-social	• costs
• carrying capacity	• morality/value system	attitudes/practices	• efficiency
• regeneration	• self-conception	• cooperation/partnership	• economic cooperation
• resilience	• common good	• manners	• sufficiency
• self-regulation	• zeitgeist	• harmony/peace	• circular economy
• fitness	• world view	• collective identity	• economic risk management
• ascendency	• time perception	• work	• social compatibility
• cyclicity	• enculturation	• humanization of work	• substitution
	• family/cultural group	• equal opportunity	
	• homeland	• self-help	
	• cultural identity	• solidarity	
	• tradition	• community work	
	• innovation	• social security	
	• differentiation	• social mobility	
	• cultural diversity	• sufficiency	
	• adaptive lifestyle	• structural nonviolence	
	• freedom (of rational action)		
	• institutional freedom		
	• social freedom/liberty		

Environmental Communication for Sustainable Development

Manfred Oepen

When preparing a position paper on Environmental Communication submitted to the Development Assistance Committee of the OECD, the editor came across two controversial books which seemed to highlight the dilemma of Environmental Communication for sustainable development (de Haan/Kuckartz 1996, Maxeiner/Miersch 1996).

The first one is an analysis of more than 1oo German studies on environmental awareness, in which de Haan and Kuckartz conclude that there is no scientifically proven relation between environmental knowledge and attitudes on the one hand and behavior on the other . Their academic colleagues in English-speaking countries (e.g. Hines/Hungerford/Tomera 1987 or Dunlap 1993) confirm this bottom line. This must come as a bitter statement for any educator or communicator active in environmental behavior change. After all, do not most of them safely assume that the simple enlightenment idea will work, i.e. that knowledge will transform into the "right" awareness and, ultimately, behavior? Do not most of them firmly believe in the triumph of reason over ignorance?

The second book - published by Maxeiner and Miersch, former editors of "Natur", one of the oldest and most influential environmental magazines in Germany - is on how deliberately negative journalism and research studies contributed to a whole era of "ecopessimism" in Germany. Positive environmental trends born out of creative, rational and responsible efforts have been downplayed –examples include "green" production lines by major industries, improvements in water and air quality or the return of storks and even wolves. The two authors argue that such trends are made possible because of successful environmental awareness-raising, education and communication. "Green thinking" has become so common place that unjustified ecopessimism may well backfire, as many people tend not to believe in end-of-the-world scenarios any more. Again, similar phenomena can be observed in other civil societies (e.g. Foreman 1991 or Kabou 1993).

The two publications put Environmental Communication between two extremes - an academic "no impact at all" and a pragmatic "taken for granted". This reveals a major dilemma - namely how to prove the value and impact of communication despite the difficulties entailed in evaluating effects and deducing the cost-benefit ratio.

Environmental Communication and Decision-makers

Maybe it is this "impossible image" of sustainable development which makes policy and decision- makers hesitate to invest in Environmental Communication, or EnvCom as it is often called. Despite the research-supported and widely acknowledged impact that communication has on social change, donor agencies rarely integrate EnvCom as a strategic tool in development cooperation programs, and do not fully recognize it as a professional field. There is a discrepancy between what decision-makers express about the crucial role and importance of communication and what they are actually willing to allocate in terms of financial and other resources. A 1994 survey among high-level UN policy-makers revealed that this is true for most agencies in respect of development communication in general (Fraser 1994). Recently, the Director General of IUCN confirmed the same regarding EnvCom in particular (IUCN 1997).

Environmental Programs often show poor results, prove to be ineffective, unsustainable and insignificant (IUCN 1997):

- In Michigan, USA, public funds and a lot of time have been wasted on an environmental policy promoting a wood-fired power plant for energy conservation, as it has been rejected by the stakeholders.
- A World Bank consultant was almost killed in Thailand when people found they were not properly involved in an EIA public "consultation" for the Yom River Dam.
- An attempt to set up a nuclear waste treatment plant led people in Russia to damage it for fear of an accident like Chernobyl.
- In the Netherlands, a campaign to return batteries for safe disposal failed when people turned in so many that the recycling system could not cope with them.
- When the IUCN Species Survival Commission re-introduced crocodiles bred at high cost in Venezuela, local people immediately killed the animals that they had tried to get rid of before.
- Cleaning the Ganges with scavenger turtles seemed a good idea in India until river communities killed them for fear of attacks on humans.
- A Mangrove Forest Protection Program in Pakistan never got off the ground as local fishermen distrusted the IUCN organizers, the urban experts and "public relations" they brought in.

The reasons for such failures are often rooted in environmentalists' beliefs that scientific facts and ecological concerns are convincing on their own. But what people perceive is influenced by emotions and socialization, more than by reason and knowledge. Hence, slogans or normative appeals like "Save Water!" will rarely be effective, meaning - change practices. Planning a systematic and holistic communication strategy taking due account of people's perceptions and felt needs is rarely considered, even though it determines the success or failure of a given program. Instead, many environmental initiatives work in isolation from assigned problem-solving institutions such as extension services, or neglect staff training in related skills. Inflated expectations that the "cognitive power" of the word and the image alone will solve a given problem in a "Said - Done!" short-cut (see PART 3) often disregard communication barriers. Communication activities, at best, are conducted on an ad hoc and sporadic basis, mainly using top-down mass media while neglecting public participation in community media.

Often, the "homework" of each and every project or program, namely empirical social and other research, especially KAP analyses (Knowledge, Attitude, Practice) is neglected. The same applies to multidisciplinary, sociocultural evaluation studies and process documentation. Needs-based, demand-driven and participatory methods are poorly applied while vertical, mass media-oriented approaches dominate. Conflicts of interest are fought by stakeholders, not negotiated by "shareholders". Approaches based on confrontation lead to one-way information dissemination disregarding understanding, instead of relying on two-way communication towards "shared meaning" and "win-win" situations.

Many decision-makers do not know how to incorporate a communication strategy in their environmental policy or project life cycles and, hence, are not willing to invest in this. The implementation dimension of initiatives is overemphasized in comparison to capacity-building and human resource development. Often, outsourcing of communication aspects of a program leads to a poor "institutional memory', especially if foreign consultants execute a program without a sound exit strategy. Media campaigns mostly target the individual and try to foster behavior change. However, communication derives from the Latin term "communis facere", i.e. to establish a community. This is a process that requires organizational and social, technical and economic, diagnostic and evaluative skills - skills derived from dialogue and interaction with peers, neighbors and friends. The net result of such deficiencies is an enormous loss in efficiency, sustainability - and financial resources.

Myths and Realities about Environmental Communication

Communication is mistrusted by "hard" technical fields as it deals with "unpredictable" human behavior. However, EnvCom is a professional field well rooted and drawing from experience in development communication, adult education, agricultural extension, family planning, community development, health and other fields. Its methods, instruments and techniques have been both theoretically founded and tested in the field. "Nobody owns EC" is a comment often heard as communication is usually not restricted to one administrative or operational level. But many organizations either maintain small communication units or establish regular contacts with related experts in extension agencies, NGOs, academia or the private sector. The impacts of communication and education are considered to be long-term and hard to evaluate. Nevertheless, FAO's Strategic Extension Campaigns - a model for the Environmental Education, Training and Communication (EETC) approaches of the UN Food and Agriculture Organization (FAO) and the World Bank (Adhikarya 1994) - provide excellent examples of how gains in knowledge, attitudes and practices of EETC beneficiaries and related cost-benefit ratios are evaluated. Also, education and communication interventions are often considered insufficiently output-oriented . But, ultimately, EnvCom aims at structural changes achieved through cooperative efforts of individuals or groups, usually in a geographically defined community. Gains and changes in knowledge, attitudes and practices are part of this process. Related targets are therefore defined from the very outset of a project within the context of a sustainable development vision.

Numerous studies prove that Environmental Communication has a positive impact on agenda-setting in a civil society (i.e. what is being discussed and, mostly, taken to be the truth) and on "low-cost" practices (e.g. recycling). Recently, also "high-cost" practices (such as giving up a car) have been tackled by EnvCom through complex sociocultural cost-benefit analyses that take life-styles, preferences and group dynamics into consideration. Research has revealed that learning to communicate and cooperate in small groups fosters constructive environmental behavior. Information input alone turned out to be irrelevant here. And communication - which is much more than information - is needed to convey a complex new paradigm such as "sustainable development" where "nature conservation" is not enough.

In a professional communication strategy, social marketing and mass media will have a role to play. However, these instruments cannot stand alone. Experience with development communication clearly indicates that you cannot "sell" ideas - in this case on the environment - like a washing machine. Changes towards sustainable development are "high cost" and, thus, have to be attractive. Social marketing, community participation, delivering benefits instead of information, and solving barriers people face instead of "educating" them was the key to success of models such as ABC - Applied Behavioral Change (see PART 3 and 5, Smith 1995). ABC helps managers to design, implement and evaluate interventions that lead to behavior change on a large scale - as needed with regard to sustainable development.

Recent Developments in the Field of Environmental Communication
Against this background, a number of development agencies have been calling for more attention to EnvCom since the mid-1990s. GreenCOM, the USAID Environmental Education and Communication Project, was created to provide technical assistance to environmental projects in order to increase and measure their impact. In 1993, GreenCOM began promoting environmentally sound policies and practices in developing countries through the application of education and communication methods (GreenCOM, 1997). The Mediterranean Environmental Technical Assistance Program (METAP) commissioned the "Handbook for Environmental Communication in Development", published by a team from Ohio State University, USA (Fortner et al 1994). The IUCN Commission on Education and Communication started assessing NGO experiences in environmental education and communication as well as people's participation in environmental policies (e.g. van Hemert et al 1994).

An international workshop on "Communicating the Environment", organized by the Deutsche Gesellschaft für Technische Zusammenarbeit (GTZ) in Bonn, Germany in late 1996 brought together a group of environmental *and* development communication specialists, unfortunately a rare occasion. A lesson the environmental specialists learned was that EnvCom can be regarded a subcategory of development communication, and that most of the latter's means, methods, instruments and techniques can successfully be applied with respect to sustainable development. The synergy effects that resulted from the GTZ workshop - in which the editors and a number of contributors to this reader participated - proved creative and productive indeed. Within two years, the network of experts from government agencies, NGOs, research units, development agencies and members of the IUCN Commission on Education and Communication

- approached OECD with a proposal for "Environmental Communication Practical Guidelines" (Env. Communication 1999) and initiated a related working group as part of the Development Assistance Committee (DAC),

- met and merged with a group of agricultural extensionists supported by the UN Food and Agriculture Organization (FAO) and the World Bank's Economic Development Institute,
- collaborated in terms of training curricula development (see PART 7),
- and elaborated this Environmental Communication reader.

GTZ, in the process, had a four-day staff training course on EnvCom designed, first implemented with WWF in 1999. The German Foundation for International Development (DSE), a "global player" in training events and methodology, had a two-week training module and manual on EnvCom elaborated by early 1998. It was used in a long-term environmental management training program in Viet Nam and urban environmental projects in Indonesia.

The aforementioned initiatives are definitely steps in the right direction. Nonetheless, in terms of professional recognition, fund allocation and training needs identification, Environmental Communication still has a long way to go. To this end, the lessons learned, successful cases, well-established methods and researched facts on EnvCom should be made more visible and transparent to policy-makers and project managers of programs related to the environment. The impact and cost-benefit ratio of communication activities should be stressed in related evaluations. Also, training curricula, both providing basic communication skills for environmental specialists and upgrading the skills and number of EnvCom professionals, should be developed and implemented. A clearinghouse for qualified EnvCom consultants, training and consulting agencies, relevant literature, etc. should be established. Guidelines and training modules concerning communication strategies in environmental programs should be developed and made available to both donor agencies and end users. Goal-oriented, systematic and well documented and publicized initiatives will no doubt provide the necessary proof of the value and impact of communication for sustainable development.

Squeezed between an academic "no impact at all" and a pragmatic "taken for granted", the field of Environmental Communication is struggling to provide proof of the value and impact of communication for sustainable development. As will be outlined in this reader, EnvCom certainly *does* have positive impacts in project cycle management, these effects can be evaluated properly and their cost-benefit ratio is a promising one.

PART 3 - Conceptual Framework

Environmental Communication in a Context
Manfred Oepen

In this section, Environmental Communication will be put in a context by offering some definitions, linking it to closely related fields and cross-referencing it to other sections of this reader. Both governmental and community-based approaches are discussed in a non-exclusive manner. In any policy or project life cycle, concepts, technologies and skills related to environmental sustainability need to be communicated to policy-makers, opinion leaders, strategic groups or the public at large. Because of the role of social learning, public participation and practice change in sustainable development, special attention is given to a community-based approach to EnvCom.

What Environmental Communication is all about

Environmental Communication is the planned and strategic use of communication processes and media products to support effective policy-making, public participation and project implementation geared towards environmental sustainability. EnvCom is a two-way social interaction process enabling the people concerned to understand key environmental factors and their interdependencies, and to act upon related problems in a competent way. As such, EnvCom aims not so much at information dissemination but at a shared vision of a sustainable future and at capacity-building in social groups to solve or prevent environmental problems. Embedded in a well-defined communication strategy, EnvCom makes efficient use of methods, instruments and techniques well established in development communication, adult education, social marketing, agricultural extension, community health etc.

In the current debate on sustainable development, communication and education as the driving forces behind environmental learning processes have an impact on at least two levels:

1 environmental awareness is determined by cultural contexts, visions, life-styles and value judgments - all of which are learned through communication,
2 criteria and options for decisions regarding sustainable practices are a result of public discourse and transparently communicated alternatives,

Ultimately, so the conventional wisdom goes to date, sustainable development cannot be based on behavioral manipulation but relies on reflection and plurality which will help civil society to develop adequate skills to overcome the ecological crisis (see for example de Haan 1997).

In an academic assessment of "Environmental Communication in Development" commissioned by the Mediterranean Environment Technical Assistance Program (METAP), scholars from Ohio State University (Fortner et al 1994) argue that EnvCom stands apart from other fields of communication in terms of the

Complexity of Environmental Issues
EnvCom deals with science, economics, law, business management, politics and human behavior, and their many trade-offs and interactions in a holistic way.
Comprehension Gap
What the lay public knows and understands about the technical dimensions of the environment differs widely from the know-how of experts.
Personal Impacts
As "nature" is often associated with traditional beliefs and sociocultural norms, EnvCom triggers reactions in nonrational, e.g. emotional and spiritual dimensions of human behavior and practices.
Risk Element
Risks are a frequent factor in EnvCom, especially as distinctions between passive/uncontrollable or active/voluntary actions are concerned.
Large-scale
Environmental interventions, e.g. in watershed management, often require coordinated moves of large populations which, in communication terms, cannot be facilitated by individualistic or small-group approaches.

A group of international environmental and communication specialists, brought together by the Deutsche Gesellschaft für Technische Zusammenarbeit (GTZ) GmbH concluded that EnvCom is very closely related to development communication, and that most of the latter's means, methods, instruments and techniques can successfully be applied with respect to sustainable development (GTZ-PVI 1997). The various fields of action directly related to environmental education and communication were defined earlier as follows (GTZ 1994):

Hence, EnvCom, is closely related to non-formal environmental education (NFEE) and social learning processes, which focus on people, processes and participation. In the long-term perspective both EnvCom and NFEE build on the factual knowledge of formal education, to which vocational pre-service

Formal fields of action	Non-formal fields of action
Formal education	**Non-formal education**
Provides factual knowledge on the physical, chemical and biological links between complex ecological systems and their reaction to human interventions on the local, regional and global level.	fosters social learning processes concerning procedural knowledge, values and social-communicative as well as technical-scientific skills that facilitate the change of norms and practices towards sustainable development through problem-solving action.
Vocational training	**Awareness-raising**
Both pre-service and in-service, fosters advanced qualification for sustainable development planning, implementation and monitoring of "green" products and production, and the improvement and consolidation of related curricula in all professions.	is based on the notion that experiential learning and spiritual and intuitive perception can trigger an emotional involvement. There are strong connections to non-formal education and communication. In practice this approach is often used rather unsystematically.

Communication

Is understood as a dialogue that enables the people concerned to understand key environmental factors and their interdependencies, and to act upon related problems in a competent way. Unfortunately, communication is often misconstrued as a "hypodermic needle by means of which information can be injected, while it is actually a transmission belt between information and action.

and in-service training are related . Because these fields are so closely related and draw largely from the same methods and instruments, many scholars tend to refer to EETC - Environmental Education, Training and Communication. Ronny Adhikaryas contribution in this PART will deal with this in greater detail. This holistic and integrated understanding of the otherwise limited concepts of environmental education and communication may well be the only way to tackle the complexity of the social, economic and ecological dimensions which sustainable development necessarily encompasses. Furthermore, based on Adhikarya's assertions, this may be the most effective way of "mainstreaming the environment", by building strategic alliances with relevant sectoral institutions concerned with for instance population, health or agriculture and by sharing information on environment issues through existing" communication channels which have a large and regular clientele, in an institutionalized and sustainable manner (Adhikarya 1996)

Communicating the environment can basically be achieved in two ways which are non-exclusive and in ideal cases may even support each other - through central or local government organizations, and through community-based or non-governmental initiatives. This is reflected in the case studies in PART 5 as well as in the contributions in this PART. While the "Role of Communication as a Policy Tool" by van Woerkoem, Hesselink, Gomis and Goldstein proceeds from a top-down/center government perspective, the section on community-based environmental communication presented below focusses on a bottom-up/periphery option, and the text by Hamacher on mediation and conflict management somewhat bridges these positions.

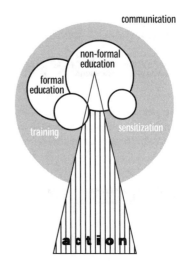

**Environmental
Action Tree**

relating communication to
formal + non-formal education,
sensitization, vocational-training

Environmental Communication in Policy-making and Project Management

Environmental Communication is a management tool, like the chain on a bicycle. The bike won't move without it but the chain cannot move on its own. Similarly, EnvCom transforms the power generated by policy-makers and project managers into action. Hence, EnvCom is the missing link between the subject matter of environmental issues and the related sociopolitical processes of policy-making and public participation. It works best in combination with other instruments such as economic incentives, laws and regulations or sectoral planning. EnvCom bridges "hard" technical know-how and "soft" action-oriented behavior change, i.e. scientific agreement and social agreement on a given environmental issue. Its high public participation potential is indispensable for the acceptance, credibility and sustainability of environmental programs.

The function of communication as a "product of government" is analyzed in greater detail in the subsequent contribution by van Woerkoem et al in this PART. They outline how, in a policy life cycle, EnvCom plays a crucial role at all stages: identifying the issue, formulating the policy, implementing the policy and management and control. The same holds true for project management. Problem identification, agenda-setting, project goal formulation, implementation, evaluation, management and control etc. cannot do without properly defined communication support. Concepts, technologies and skills related to environmental sustainability need to be communicated to project managers, opinion leaders, strategic groups or the public at large. Breaking down complex information into understandable elements and putting those on the agenda in a socioculturally relevant and economically feasible way to different audiences, is a prerequisite for consensus-building and change in any civil society. Hence, EnvCom is part of the lifeblood of enlightened and transparent decision- making and effective action towards environmental sustainability.

It is essential for a project planner to realize that different actors are involved at each stage, and that each actor has different perceptions and interests. The potential contributions of communication are related to the various stages of the project life cycle. During the *identification* phase, the role of the project manager increases until it reaches a peak when the problem at hand is *formulated* and tackled. Public awareness of the problem decreases when solutions are *implemented* but still needs to be maintained and *controlled*. During all these stages, communication plays a continuous, yet different role - as indicated below.

Identification Regular opinion/attitude surveys - media content analysis - continuous networking with NGOs, consumer groups - regular meetings with interest groups

Formulation KAP surveys - integrating communication in the mix of policy instruments - design of communication strategy - communication with those involved

Implementation Communication as an independent and as a complementary instrument - information on other instruments (laws, incentives etc.) - M&E through qualitative research

Control Regular public information - reporting on changes in policy design and implementation - up-dated opinion/attitude surveys.

Understanding where the project is in its progression from identification, formulation, implementation to management is an essential basis for determining which communication instruments should be used. Knowing "what" should be changed has to be combined with "how" change should be brought about. Methods of EnvCom that fulfill different functions in changing practices (**K**nowledge - **A**ttitude - **P**ractice) can roughly be assigned to the two major phases of the project life cycle - planning and implementation.

Methods	Planning	Implementation
Knowledge information dissemination	**Agenda-setting+information** on • the environmental problem • potential solutions offered by relevant organizations • actual participation options • potential activities • surveys on acceptance of planned activities	**Information on** • actual state of the environmental problem • ongoing activities and their success • opportunities for participation • necessary changes of practices, incentives and sanctions • surveys on acceptance and success of activities and changes of practice
Attitude experiential learning	**Applicable at any given time** • planning support • information and opinion surveys • sensitization for changes of practices	
Practice action orientation	**Involving relevant actors** in developing solutions to • the program at large and particular activities • the communication strategy as an integral component • networking, establishing lasting cooperation	**Involving relevant actors in** • monitoring and process-oriented modifications to ongoing activities ("action-learning") • continuation of participation started at all levels of the program and its communication strategy • evaluating completed activities • deriving "lessons learned" for future expansion

This can be exemplified by means of the role of different communication instruments in the various phases of Protected Area System Planning for a marine conservation program in Indonesia.

Phase in Park Management	Methods of Communication
1 - **Preparation**	• Personal visits to the park with stakeholders to qualitatively assess the extent of the problem for the people affected • Qualitative knowledge/attitude/practice (,KAP') surveys • Contact with non-governmental or community-based organizations which will implement the EnvCom strategy • Basic information material on the park environment and the necessity of conserving the area to be distributed to relevant groups • Regular briefings, interviews and meetings with interest groups in order to give updates on the conservation process
2 – **Composition**	• Quantitative KAP surveys • Integrating communication in the mix of policy instruments • Design of communication strategy • Extension to and communication with intended stakeholders and beneficiaries
3 - **Implementation**	• Communication to raise awareness of conservation issues among key groups of the local population • Inform groups on the use of other management instruments (new legislation, subsidies, alternative technology)
4 - **Maintenance**	• M&E through qualitative research • Continued public information • Regular opinion/attitude surveys

Environmental Communication is not a Magic "Said - Done!" Tool

Information is not the "missing link" between a problem and a solution. Inflated expectations that the "cognitive power" of the word and the image alone will solve a given problem in a "Said - Done!" short-cut often disregard communication barriers (see illustration). Here, Liebig's Law can be applied: the yield is related to the one indispensable nutrient (light, water, fertilizers etc.) which is available in the smallest amount. In other words - if a flower doesn't see the light, you may water it as much as you want, it will not grow. Applying this law to the growth of an environment or development program, even the most sophisticated communication strategy will not solve a problem if a minimum level of economic resources, social organization and political bargaining power is lacking. For example, a rural woman selling vegetables at a market place is sitting under a health extension poster "Eat more Vegetables, and your Health will Improve!". She has grown the cabbage and the onions, and knows exactly how nutritious they are. However, she is forced to sell her produce for cash in order to pay the school fees for her children and the kerosene for her stove. The information input is in vain.

Information is the basis of any effective public education and communication program. Making information available to society at large is a public service, helpful for decision-making and perhaps a statutory obligation. However, providing environmental information is often passive, and depends on people actively seeking it. Therefore, the information does not necessarily reach or interest people

Said is not heard

Heard is not understood

Understood is not accepted

And accepted is not yet **Done**

There is more between a problem and a solution than "Said - Done!". Many people know this from anti-smoking campaigns - messages and information about cancer are abundant and taken in ("heard" and "understood"), the doctors are believed so that attitudes towards smoking are negative ("agreed"), and still, giving up smoking is difficult to achieve and to maintain ("done"). Germany provides another example: 80% of the people feel negatively affected by traffic. Due to a high dosage of daily media coverage, they know, understand and agree that driving cars is environmentally "detrimental". And still, almost all of them own a car.

who are causing a threat to biodiversity, for example. Traditional thinking in the field of public education has been that making people more knowledgeable about the environment and its problems can change their behavior. This was based on the assumption that increased knowledge directly leads to a greater awareness of the environment and to environmentally friendly behavior. However, research does not support this assumption. The results of some studies even suggest that greater knowledge about environmental problems can lead to a state of general anxiety, a denial of the problems or a complete refusal to think about environmental problems. Research indicates that in-depth knowledge about and personal feelings towards environmental issues must be accompanied by skills with respect to appropriate forms of action which promote self-confidence and participation in environmental issue-solving.

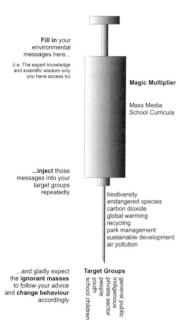

Vertical Model of Environmental Education and Communication

Many planners tend to think that producing posters and video films or launching a mass media "campaign" is a solution to problems rooted in environmentally unsustainable practices. Such isolated ad hoc initiatives that are not integrated into a comprehensive communication strategy, will just cause inflated expectations in rational appeals and the cognitive dimension of messages. In much the same way, environmental education has concentrated on schoolchildren. Where schools are regarded more a public space than an institution, students, teachers and parents may meet and exchange innovations and plan for specific environmental action in their community.

The aforementioned expectations concerning the "cognitive power" of the word and the image, mostly based on attitude-behavior models from applied psychology, can often be observed in environmental scientists and planners, too. They fill in their environmental messages - i.e. the expert knowledge and scientific wisdom to which only they have access - in what they consider magic multipliers, such as mass media or school curricula. They subsequently inject

49

their supposedly ignorant target groups with those messages – like with a "hyperdermic needle" – and optimistically expect the "floating masses" to follow the experts' advice and change their behavior accordingly (see illustration).

Key terms, here, are products, messages, issues and technologies. There are many reasons why these models have not been successful:

1 the relationship between environmental awareness and behavior is very weak,
2 behavioral change rarely results in lasting and significant effects with respect to environmental quality,
3 expected changes in behavior as an objective can be counterproductive as they often violate individual self-determination and self-motivation,
4 specified behavior changes tend to fragmentize environmental knowledge and to isolate environmental degradation as personal problems,
5 as such, the Attitude - Behavior - Model often produces docility instead of critical thinking, empowerment and environmental competence.

Moreover, the models often fail to comply with some simple pedagogic rules (Weltner 1993), which tend to be neglected:

Pick learners up where they stand
If new concepts, especially complex and controversial ones, have no connection with what people know and believe, and with how they see things, the innovation tends to be rejected. Therefore, communicators and educators should make great efforts to understand and master the language, terms, rationales, mental images, historical and social context of the people with whom they are working. Only when these factors are understood, can they serve as a basis for new concepts and careful linking.

Didactics won't work - learners imitate others
Action is stronger than words. Therefore, communicators or educators should not only preach but practice, they should always practice what they preach and, if possible, they should have some powerful role models - like sports heroes or other idols - to increase their impact.

You can only learn with your own head
You cannot learn "for" someone else. People have a tendency not to listen or learn if they don't have an interest in the issue at stake. This is why media produced for people are less effective than media produced by or with the

people concerned. Learning *and* teaching is based on active listening which requires understanding, i.e. reconstructing and interpreting the subject matter in question by means of one's own terms, experiences and perceptions. Educators, therefore, should always stimulate this active listening and be able to shift to the learner's level of concepts and language.

These observations stress one of the most important lessons from education and communication research - that most people are very smart and have excellent reasons for doing things which seem self-destructive to outsiders, sometimes virtually "sawing off the branch they are sitting on" as a German saying goes. To help people protect the environment, EnvCom planners should understand their world view and offer "benefits" they care about, reduce "barriers" they worry about, and promote attractive alternatives to the harmful practices they are engaged in (for more details see for instance . the section on "social marketing" in PART 4).

Very often, people engage in environmentally detrimental practices because they have no other economic choice, or the incentives to do so are strong. In such cases, people are often very aware of what they are doing so that "awareness-raising" is unlikely to be effective. The approach of communication and educational programs, here, should be to engage relevant stakeholders in problem-solving to determine what personal benefits can be emphasized and what changes are needed in the social and economic system to support effective action. Without structural changes, education or communication programs designed to stop such practices are in vain. Therefore bottom-up communication should be directed towards influencing policy and direct interventions, and to having the appropriate support put in place. Top-down communication can be used to inform people of the measures and to create a situation where "old habits" are no longer socially acceptable. This is why communication as a management tool is strongest when it works in strategic combination with measures to lower barriers to change. Hence, communication and education instruments should have equal standing with economic and legal policy instruments, and should be used in strategic combinations with them.

Even where environmental information disseminated by the mass media is accepted, it will, in general, only cause short-term knowledge gains and attitude changes, e.g. related to "biodiversity". Taking this for "sustainable development", however, would be fatal and shortsighted. The acceptance and effectiveness of information messages rely less on content or format but on sociocultural context and the source's credibility. Here, interpersonal networks are more binding and more relevant for social transformation than overburdened public services or anonymous mass media. This phenomenon can be observed even more clearly in peasant societies, which are based on one dominant principle - subsistence insurance, i.e. economic reasoning, conformity to homogenous cultural values, patterns of production, social life and communication to minimize the risk of failure and loss. Changes in environmentally relevant practices that generate sustainable development depend strongly on the association of people, on increased availability of options for action, on the organization of networks - as clearly stated and called for in the Local Agenda 21, for example. Nevertheless, networks, groups and communities as a driving force in development have been neglected by individual-psychological modernization and media theories which also codetermine the mainstream of environmental institutions' and planners' thinking despite many findings in theory and practice (Fremerey 1981, Sülzer 1980). Similarly to the "diffusion of innovation" approach that made use of network analysis (Rogers 1976), attempts to induce environmental changes by information dissemination often encounter the "information effects gap" (Singhi/Mody 1976): information is absorbed along economic and political determinants which favor the well-off

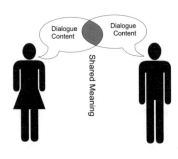

and disadvantage the non-privileged. Therefore, information alone is very often *not* the missing link between a problem and a solution; dialogue is.

Key terms in this model are process, feedback, dialogue and people. Its horizontal, community-based approaches can be complemented very well with vertical support systems involving the mass media. Therefore, the limitations of mainstream media-based information and communication strategies should by no means deny the role of mass media in sustainable development. However, sustainable development defined not as a cluster of benefits handed down to the people, but a process by which the people gain greater mastery over their destiny in a socially, economically and ecologically sustainable way cannot do without social interaction. Given the complexity and effects of sociopolitical contexts, mass media, especially in the so-called Third World countries, usually prevent social interaction, while small, group or community media usually instigate it. Therefore, it cannot be stressed enough that EnvCom which is meaningful to innovative and democratic sustainable change, at least in crucial stages, cannot

Horizontal Model of
Environmental Education
and Communication

dispense with interpersonal communication in relatively small social units, e.g. groups, networks, communities. Even though this may be obvious and generally agreed upon, it must be restated, because mass media and information technologies radiate such a "magic" and overpowering fascination. Hence, a communication model is needed "that puts people not at one end as sources and at the other as receivers but right smack in the middle ... With such a model perhaps the media will cease to look larger than life and be perceived as first-class tools, but tools, nonetheless, to be used by development and communication planners but also by the people on whose behalf plans are made" (Quebral 1988). In the non-ideological and pragmatic approach to EnvCom advocated in this reader both vertical mass media and horizontal community media play decisive roles.

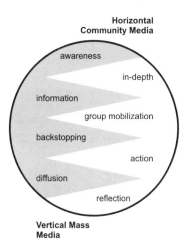

Complementarity of Vertical Mass Media and Horizontal Community Media

Because of the decisive role of social learning, public participation, action orientation and, ultimately, practice change in sustainable development, special attention should be given here to a community-based approach to EnvCom.

Community-based Environmental Communication

Environmental interventions are very often initiated by organizations and policy-makers from outside the community. Many well-intentioned communication projects of this sort turn out to be efforts at manipulation that result in little or no participation by the people concerned. Instead, dealing with cross-cultural communication and local media should focus more on how to listen than on how to talk (Ramirez 1997). Communication begins by learning to learn about existing knowledge and hopes. Listening requires skill and respect, and the best communicators tend to be those who have trained themselves to learn and derive meaning from different media: from the elders' anecdotes and oral history, from the artists' symbols, their songs and poetry, and from traditional theater and puppetry. Before effort and resources are dedicated to explaining outsiders' proposals, conservation workers must first learn about local perspectives, indigenous knowledge and people's hopes for the future. This task is the essence and starting point for cross-cultural communication to take place in the context of a rehabilitation or conservation program. The field of communication lends a hand to environmental activists and planners by making the task of listening respectable on a professional level and rewarding. Communication is about bridging understanding. This is done by exchanging messages to create meaning and to enrich the knowledge base of communities to help them face change.

It has been noted so far that vertical models of communication (sender - media - receiver) and related centrally planned development strategies alone have proved incapable of solving today's acute environmental problems. Their basic

	Top-Center Models	Bottom-Periphery Models
Goals	behavior change, social engineering, educational/professional competence, security of resources, government authority	self-expression, conscientization, emancipatory action, social+ political competence, structural change, citizen power
Interests	central control, transfer of values + norms, predictibility of behavior, property rights, cultural and artistic direction	local and regional autonomy, democratization of civil society institutions, increase of countervailing power,
Content	standardized messages, central instructions + guidelines	local relevance, problems and solutions, action mobilization, self-help motivation
Sender	central services and decision-makers, professionals	local or community groups, NGOs, grassroots organizations
Receiver	masses, peripheral, passive, ignorant,	like senders and their clientele, interacts with senders
Diffusion	linear, vertical, top-down, multi-media, political socialization, information control	process and dialogue-oriented, horizontal, bottom-up, political participation, information access
Form	limited projects, campaigns, slogans, power of the word	integrated in community development, self-help organizations, long-term, power of action
Media	hightech, mass media, capital and technology-oriented, media FOR 'target groups'	group, low-cost, traditional, community media, media BY and WITH beneficiaries
Support	training of professionals, media and information technology infrastructure	motivator training, mediation, networking, documentation, complementary action in community development, self-help

Communication Model Continuum

problem is that nothing goes without changes in practice and that this change requires social action facilitated on a horizontal level, mostly by means of interpersonal and group communication. Horizontal models of communication (communicator - dialogue - communicator) alone also proved to have limited impact . Here, nothing goes without lobbying for one's interests in the political, economic and social arena through alliances with other social institutions at various levels and a general increase in "bargaining power" via communicative, social and political competence. For pragmatic, historical and leverage reasons, many "intermediary institutions" such as NGOs, cooperatives or church organizations which often bridge the gap between community groups and "superstructure institutions" such as banks, research agencies, government services or mass media underwent a general reorientation - from "information diffusion" *to and for* the people to "information seeking" *by and with* the people. Here, problem- and practice-related information is generated through local or regional community processes and used as inputs for existing media networks horizontally and vertically, both - bottom-up to inform central decision-makers and top-down to inform community groups in different places (Sülzer 1980).

In the light of such considerations, an integrated approach to community communication for sustainable development may be useful which could pragmatically overcome deficiencies of both purely vertical and purely horizontal models, while building on their strengths. Community communication does, in fact, link and facilitate networks between

- different societal levels, e.g. community, intermediary, superstructure level,
- different social groups or organizations on those levels, e.g. self-help groups or local elites at the community level, NGOs, environmental or church organizations at the intermediary level and mass media, research institutions or extension services at the superstructure level,
- different communication systems, e.g. mass, traditional or interpersonal communication,
- different media - from folk or street theater, slide-shows or video to radio or TV - on various levels.

Community communication is defined here "as a process of horizontal and vertical social interaction and networking through media regularly produced, managed and controlled by or in close cooperation between people at the community level and at other levels of society who share a sociopolitical commitment towards a democratic society of countervailing powers. As the people participate in this process as planners, producers and performers, the media become informing, educating and entertaining tools that would also make non-privileged and marginalized people think and speak for themselves, not an exercise in persuasion or power. In such a process, the entry points for communication interventions should be sought in the communities' learning methods, cultural expressions and media forms. Community communication, hence, describes an exchange of views rather than a transmission of information from one source to another" (Oepen 1995:49). This understanding and also the three key concepts of community communication - access, participation and self-management are designed to minimize possibilities of oppression and power abuse (Community Media 1983):

- **Access**, at the level of choice, means the right to communicate, the availability of a wider range of materials chosen by the public instead of production organizations, and the transmission of materials requested by the public. At the level of feedback, access implies interaction between receivers and producers of messages, and active participation of the audience during program planning and production.
- **Participation**, at the planning level, is the public's right to contribute to the formulation of plans, policies, objectives, principles and programs of media immediately affecting given communities. At the decision-making level, it implies public involvement in programming, management, financing etc. of those media. At the production level, it opens opportunities for the public to produce programs and have access to professional help in making technical facilities and resources available.
- **Self-management** is the most advanced form of participation as the public, in this case, would fully manage and produce community media and determine its contents, goals and policies.
- The author would like to add a fourth one here: **accountability**, i.e. an organizational form of the groups, institutions and media involved in community communication which allows them to be answerable to their members for their actions.

The understanding of communication and key concepts outlined above coincide fully with the requirements voiced by many environmentalists who seek social sustainability in conservation, and regard public participation therein a conditio sine qua non for sustainable development as such (IUCN 1997). The "environmental equivalent" of the community-based communication approach is "Primary environmental care".

Community-based EnvCom tends to overcome the old discriminations between mass media and popular media, top-down and bottom-up approaches without neglecting their different roles and functions in a given society. Only a case-by-case analysis that starts from a local or regional environmental problem, is culture-specific and takes political-economic as well as ecological factors into consideration will answer the crucial questions: "Who determines what sustainable development should be?", "What do the media do to support that kind of development?" and, hence, "Do particular media or communication strategies under the given circumstances meet the above-mentioned criteria?"

The *"horizontal or bottom-up is good'"* versus *"vertical or top-down is bad"* dichotomy does not help much in answering these questions as it rarely exists or can be treated so easily in reality. For example, the local elite in a village that plays a role in park management works mostly on a horizontal axis and may act on behalf of village interests vis-à-vis "central" forces ("good") while defending their own privileges against other community groups ("bad"). Environmental NGOs work mostly on a vertical axis as an outside catalyst trying to facilitate self-determined community organizing and grassroots' voices in a bottom-up manner vis-à-vis "central" forces ("good"), while they can be paternalistic and top-heavy in their community training or dependent themselves on governments' political goodwill or foreign donors ("bad"). Many environmental organizations and decision-makers have realized that sustainable development is not a zero-sum matter, e.g. an environmentally friendly innovation benefiting the poor must not necessarily be at the expense of the rich. Furthermore, participatory approaches ultimately not only serve social harmony but increase the quality of extension and are more cost-effective than programs implemented in a top-heavy fashion.

Moreover, the *"small or traditional is good"* versus *"big or modern media"* dichotomy does not lead very far as borderlines have been blurred by merging technologies and the wide availability of relatively cheap, yet sophisticated media. The problem with their use, as stated previously, is often not the technology but the political will. Options other than central, vertical and linear communication strategies should be feasible not despite but because of "modern" media: VHSC/Video 8 mobile units or desk-top publishing clearly indicate technically possible and economically feasible trends towards integrated media systems in decentralized units such as local radio, photo labs, video studios, small publishing houses etc. Whether or not they are used on the basis of a shared commitment to democratic social transformation and ecological concerns can be evaluated on the basis of criteria such as access, participation, control and self-reliance which the media, individuals and groups involved should support. For example, local radio or TV - such as Radio Veritas in the

Primary environmental care - PEC

PEC is a process by which local communities — with varying degrees of external support — organize themselves and strengthen, enrich and apply their means and capacities (know-how, technologies and practices) for the care of their environment while simultaneously satisfying their needs. The intelligence, experience, interests and priorities of people and communities, and their willingness to work together for common objectives, are what PEC is all about. In synthesis, PEC integrates three objectives:

Meeting local needs means that people can maintain, produce or gain access to the goods and services (food, shelter, income, health care, education, transportation, etc.) necessary for life, health and well-being.

Protecting the local environment means different activities under different conditions (e.g. eliminating a fire hazard, cleaning and protecting a watershed, preventing flooding, halting unsustainable extraction of timber from a local forest, improving tilling practices to protect topsoil, restoring a degraded communal building, leaving the habitat of wildlife undisturbed, etc.).

Empowering local communities means that communities, groups and individuals get more control over the factors influencing their lives. This usually involves several stages, in which people discuss and identify their common problems and opportunities and then organize and take action in partnership with others. Securing tenure for the natural resources protected by the work of local people is a most important element of the empowerment process, and is essential for sustainability. With security of tenure, in fact, the long-term economic interests of people tend to merge with the long-term "interests" of the environment (Borrini-Feyerabend 1997b).

Philippines or Kheda TV in India - are "in" as long as they allow for access and participation while "folk media" - like the puppet shows for family planning propaganda used by the Ministry of Information in India or Indonesia - are "out" if they transmit centrally programmed messages to passive receivers.

To ensure that a community-based EnvCom approach is not merely idealistic or lipservice, certain minimum indicators that its "cornerstones" are accomplished must be established. Such indicators could be laid down by all partners involved in a more or less formalized temporary "social contract" (for more details see the initial stages of the EnvCom Strategy in PART 4). These "Terms of Reference" of the partners in dialogue should also include a minimum definition of *sustainable development* and *democracy*. Here, a "relativity of cultural relativity" should be applied, leaving room for cultural diversity of "sustainable development" while ruling out the interpretations of "democracy" by the Hitlers, Idi Amins, Dengs and Pinochets of this world (Preiswerk 1978). The "contract" should be evaluated thoroughly by means of participatory empirical research as to the degree of participation, access, accountability or action orientation. In evaluations, it is crucial to consider that the communication process both on the horizontal and the vertical axis on one hand and the media product on the other hand are equally important. Even small increases in dialogue between people involved in a development situation may result in unexpected dynamics and effects - which may not show in the media products but could be observed in the communication processes. Also, the professional quality of or the degree of participation in the media product may not necessarily reflect the spirit, effects and consequences of the communication processes which were triggered by the related environmental program.

What counts is that the "social contract" and a minimum understanding of the "cornerstones" of the community communication approach are shared by all groups and partners involved, i.e. the community members have achieved a certain degree of access, participation, self-determination and accountability in planning, implementing and evaluating related programs and media. If that is the case, it does not matter too much what kind of media are used or where the initiative for innovation originated. At the basis of a community-based EnvCom approach there will always be a number of steps and processes controlled by or carried out in close cooperation with community groups and media. Ultimately, it will not matter which other social groups or institutions from different societal levels the community groups seek affiliation with - whether an NGO on the intermediary level or an environmental research center on the superstructure level. Also, the consolidation steps of such an approach basically only repeat in cycles what was tried out earlier - possibly at ever higher qualitative levels.

Despite their often undogmatic, experimental and pluralistic nature, some common elements of community-based communication experiences and projects around the world can be summarized. The list is neither complete nor are its entries "indispensable". Instead, the list should be regarded as containing elements which have repeatedly been found favorable and constituting aspects of the approach in question. Community-based communication

- identifies, analyzes and tackles the realities and problems of specific local groups which are often characterized by subsistence management, poverty, lack of organizational capacities and of access to productive and intellectual assets,
- offers opportunities to community groups to initiate, plan, manage, produce, distribute and evaluate media for local developmental and environmental problem-solving as well as social and ecological analysis and action, narrows the gap between different community media, groups or organizations on a horizontal axis, and mass media, professionals, banks, research or government agencies on a vertical axis,
- taps the creative and innovative pool of local knowledge systems, traditional culture, learning methods, expressions and formats, and hence helps to regain and maintain cultural identity and social self-reliance,
- promotes long-term, structural change - instead of short-term, limited, sectoral projects or campaigns - through communicative, social and political competence for more and better bargaining and countervailing power of non-privileged groups in a democratic society that safeguards individual emancipation, civil rights, public control of state and private political and economic power centers, political and cultural pluralism,
- allows for participatory instruments of project monitoring and feedback which facilitate a type of "information seeking" from below that is essential for effective planning and implementation in complex development contexts, in which mediating individuals and institutions such as local leaders, churches, NGOs etc. often play an important role.

Challenges of Environmental Communication

In terms of contents and themes, EnvCom needs to

- clarify "ill-defined" concepts such as "sustainability", "nature", "participation" etc. through tentative arrangements of priorities, and
- elaborate a limited, coherent and comprehensive core curriculum of environmental education and communication, divided into "Must Know", "Should Know" and "Nice to Know".

As EnvCom is still a relatively new field, it needs considerable methodological improvements related to

- sharing complex concepts - both scientifically and politically controversial - in short and mostly random learning situations,

- making use of interactive experiential learning through games, exercises, simulations, role plays etc. instead of relying on cognitive learning only,
- sharing the vast methodological experience with colleagues in other fields, e.g. development communication, adult education, moderation and visualization techniques, agricultural extension, health and family planning education etc.,
- facilitating training of trainers in environmental education and communication, both with respect to environmental themes, and education and communication methods, instruments and techniques.

These challenges addressed, the visions and trends in future Environmental Communication should be to progress

- from one-way to two-way communication,
- from cognitive to experiential learning,
- from "school" as an institution of top-down teaching (vertical) to "school" as a meeting point for teachers, students and parents who engage in interactive dialogue (horizontal),
- from a product orientation (media, messages, curricula, training courses) to a process orientation (round tables, action competence),
- from one-dimensional and linear ecological moralization to a discourse on a variety of value judgments from competing social, economic, ethical and other points of view,
- from stakeholders to shareholders, or - from conflict positions to "shared meaning",
- from distributive to integrative negotiations,
- from isolated, factual and compartmentalized knowledge to the ability to deal with complexity, uncertainty and risk,
- from "behavior change" to self-efficacy, social, communicative and environmental competence and personal responsibility.

The Evolving Role of Communication as a Policy Tool of Governments

C. van Woerkoem, F. Hesselink, A. Gomis and W. Goldstein

Governments are quite familiar with environmental communication and use it to project themselves via publications, campaigns and television presentations. Through these means, governments seek to inform people on policy measures or policy proposals, with the intent of attaining their policy objectives.

In order to achieve effectiveness and efficiency, international conventions and policies have increasingly emphasized stakeholder participation. Stakeholder participation depends on new approaches that change the role of government in the policy process by engaging different viewpoints right from the beginning of a new policy and throughout its implementation. In this sense, communication is a new and underexplored instrument. Its various roles can be summarized as follows:

1. **Communication as a product of government** as part of the policy mix, as an instrument used together with other instruments such as regulations or taxes in order to influence attitudes and behavior in the community.
2. **Government policy as a product of communication.** The use of communication to improve policy processes in order to create a more effective policy by interaction with the stakeholders most affected.
3. **Government as communication.** The use of communication as an essential element of government, not directly oriented to a fixed policy product, but directed more generally to social problem-solving. In this case government fulfills its function in collaboration with other actors.

Communication as a Product of Government

Here, communication - along with awareness-raising, education and information - is deliberately used as one of the instruments governments can employ to try to change people's attitudes and behavior. Policy-makers need to decide how to mix these instruments with other, market-based or law-and-order instruments such subsidies, taxes, compensations, regulations or directives so as to achieve their objectives most cost-effectively.

How communication is positioned in the policy mix depends on how it is viewed relative to other instruments in determining what people will do. Communication can be used on its own where the barriers to people changing their practices are not too great, or there is already some internal motivation to do so. Communication in the case of campaigns is used especially where other instruments are too costly. Reality shows, however, that stimulating people to change has its limits when communication is used on its own.

Communication is most effective when used in combination with other instruments that lower barriers to change unwanted practices, creating conditions to encourage change such as providing infrastructure for recycling schemes, or tax incentives for insulating homes. In a supportive role, communication can stimulate public awareness about the existence and content of other instruments (such as subsidies for insulation), enhance the acceptability or relevance of these instruments and improve the efficiency of their implementation. For example, communication can let people know about regulations related to the trade of endangered species, informing them about appropriate behavior at home and abroad, as well as the implications of not complying. Communication can influence the attitude of people, making them more cooperative. However, if the regulation is unacceptable, the task of monitoring the desired behavior is quite difficult, as trade in species goes underground, species are hidden in false bottom cases, etc. How well the regulation can be enforced depends on the visibility of the behavior.

Policy Process

Relevance of a problem in public perception

In the policy process, there are typically four phases

Identifying A problem is realized and lobbied for by social groups, and a public discussion starts.

implementing

control

formulating

identifying

Project Life Cycle

Formulating Policies are formulated, research commissioned, and options for improvements are intensely deliberated.
Implementing Policies, programs and projects are implemented. The debate slows down while people affected remain informed.
Control The emphasis is on routine surveys. Decentralization and public-private partnerships may be considered for sustainability.

Policy processes are cyclic. Feedback from implementation and management sheds light on how to improve the policy and the measures used to implement it. Government is in a continuous process of developing new policies as issues are put on the agenda, or adapting policies due to political change. In each phase of the policy process, communication has different significance and roles. Traditionally, governments tend to use communication in the third phase – the implementation phase. The figure below shows the role of the government, below the curve, in the phases of the policy cycle. During the identification phase the government's role in the policy process increases gradually, reaching a peak when the policy is finally formulated. Thereafter, there is a slow decrease in the government's involvement as ideally parties other than the government take action, reflecting a higher degree of public self-regulation. These actors include local authorities, the business community, industry, NGOs as well as consumers.

Communication in the Environmental Policy Cycle

Identification of the issues

At this stage the role of communication is to place environmental issues on the agenda. Organizations in society play an important part here. The central government adopts a relatively low profile. Communication services need to listen to what people are saying so that they can identify problems promptly and pinpoint specific issues affecting the target groups of environmental policy. At this juncture, activities also involve communicating opinions, drawing attention to the issues, mobilizing support and defining themes. The methods of communication used in this phase are:

- regular opinion and attitude surveys
- mass media content analysis
- systematic and continuous networking with NGOs, interest groups and scientific institutions (public relations)
- regular briefings, interviews and meetings with interest groups and the press.

Formulating environmental policy

At this stage activities can on the one hand raise public awareness of the environmental problems, and on the other hand consult stakeholder groups. The problems to be tackled are those that legislators have accepted but for which solutions have not yet been found. At this point, the target groups are opinion leaders, decision-makers and the general public. The methods of communication are:

64

- knowledge / attitude / practice (KAP) surveys
- integrating communication in the mix of policy instruments
- design of a communication strategy
- communication / consultation with those who will be involved (public relations, stakeholder groups, focus groups)

Implementing environmental policy

The aim at this stage is to communicate information on how to proceed. The idea is to communicate the substance of the policy and the accompanying measures. At this time communication will focus on specific target groups. The methods of communication often employed are:

- information campaigns
- specific information materials
- marketing and advertising
- instruction
- education
- consultation of target groups (public relations)

Management and control

Here communication is provided as a service to sustain newly adopted attitudes and behavior. The aim is to provide information about the policy that is being pursued as well as to generate feedback on reactions to that policy. Communication may be in the form of an active service explaining complex legislation and regulations. It may also be used to announce modifications to policy instruments, for example legislation. The methods of communication are:

- monitoring and communication of results
- regular opinion and attitude surveys
- informing on changes of policy design and implementation
- education

Problems with this Approach

In the usual approach to policy-making, it is hoped that there will be a response to the stimulus of communication. An effect is desired from well-structured messages based on an analysis of pre-dispositions, such as the interests of the individual and his or her attitudes or previous knowledge. However, this underrates the factors influencing acceptance, and that is why so much can go wrong with communication conducted in this way. Problems of policy acceptance often arise from the method of policy-making, with the government setting the goal and devising a set of instruments to achieve it. Governments tend to look at the world according to their own definitions of the problems and perspectives on solutions. Perspectives outside the policy system are usually not taken into account. Powerful interest groups lobbying for social change for the environment are the ones involved in policy-making. There is generally not much consideration of the groups who will be affected by the policy.

What is overlooked is the way citizens perceive social problems, how they define and consider them, and what active role they see for themselves. Policy products tend to be created as a given solution. Then, the customary approach is to try to raise awareness and acceptance. Citizens are expected to change and adapt to the ideas of the government. Governments are confident about the way they define problems and solutions as their reports are based on scientific inquiries, and discussed at length in the political arena. It is difficult for them to concede that a totally different approach to the same problem is possible. This arises because the common person is not involved in the political process, and the government loses contact with its citizens. The result is a host of policy decisions directed at target groups which are not included in the decision-making process, leading to a low acceptability of government policy.

Another pitfall in the communication process is to overlook the fact that groups of people shape their own messages autonomously. People cannot be induced to think in another way, so that arguments and data for a certain position are not necessarily convincing. This type of communication approach neglects the relationship aspect: People design their own messages using their own perceptions of the situation, or indeed of nature itself. They label information from the government as emanating from an "outside group", which does not have to be taken seriously, and hence do not learn anything about what is being said. This lesson is clearly demonstrated by the Dutch de Peel case study in PART 5. An important prerequisite for communication is the credibility of the sender. People must be able to trust the intentions of the government and that it will deal with their interests in a responsible way. This credibility is not only shaped by communication, but also by the *government function*. Moreover, the historical dimension is forgotten in communication. A government's message is part of a series of communication actions over a long period of time, delivered by different sectoral areas. All merge in the mind of the individual as part of the total impression produced by government over time.

Another problem arises when communication is not considered throughout the policy process and brought in after the plan has been developed, consequently necessitating the application of a DAD model: "Decide, Announce and Defend". While the communication is informative and motivational, it does not create the desired acceptance. Top-down campaigns are typical of this instrumental approach, neglecting the cultural factor, i.e. that people adopt the rationality or perspectives of the group to which they belong. This view overlooks the fact that people change as a result of discussion about issues they consider important.

Communication specialists are not only responsible for the effectiveness of communication as a sole instrument, but also for the effective implementation of other instruments. Therefore, they must under-stand the strengths and weaknesses of regulations, etc. and have a broad orientation. In addition, non-communication instruments have a communication impact. A law is a message. If the government declares a landscape protected, it restricts farmers' activities. Farmers are likely to act negatively since the law communicates a message that the farmers cannot be trusted to protect the very landscape that they have created. Analyzing the communication outcomes of non-communication instruments is thus an important task for any communication specialist.

For these reasons, the skills of a communication specialist should be incorporated in the process of policy-making from the beginning, and not be invited in only at the end. The "DAD" approach is still very common but often results in an impossible mission. The intended messages conveyed to a given community are unable to compensate for the unintended messages which are the negative communication consequences of the chosen policy mix. It is difficult to assess in which situations and in which combinations the policy mix will be effective. In fact, the distinction between different instruments is not always clear. For example, some laws operate more as communication than regulation. For reasons such as political pressure, the invisibility of the behavior, widespread deviancy and restricted enforcement capability, many laws are more symbolic than enforceable, and their impact is simply through the message communicated.

Another variable of policy success is the ability of the target group to respond practically in accordance with the desired behavior. For example, the ability of farmers to maintain current economic profits will determine their willingness to undertake organic production in agriculture. Similarly, various measures from different government agencies can have a cumulative impact on the target group, making it difficult to analyze all context variables. The selection of the instruments in practice depends not only on rational choice, but also on many factors, such as political considerations, cultural aspects, customs and experience. For example, if many lawyers are involved in solving a problem, the solution is likely to be a law.

Unfortunately, we rarely have hard facts about the effectiveness of a combination of instruments. Communicators can however help to reduce uncertainties and stimulate improved public debate on government interventions by making the reasons behind the policy decisions more transparent and by formulating alternatives. One of the problems facing biodiversity, for example, is the acceptability of government policy. Unless a sufficient level of acceptance is reached, other instruments, including financial ones, are often ineffective. This point is often neglected on the assumption that the problem of effective governing is the choice of appropriate instruments. However, the prerequisites for their application – such as an adequate level of acceptance - are often not met. If communication is not employed until a policy is approved and implementation is at hand, its potential for fostering acceptability will be limited.

Acceptability means a positive attitude towards the policy content, not a tendency to react accordingly. The latter depends on external stimuli, such as financial incentives or punishment. However, in the long run, it is difficult to guarantee a stable level of behavior through these measures alone. We need a term to denote whether people are willing to comply, regardless of such positive or negative incentives. Acceptability is multidimensional and includes the attitude to

- the problem itself, i.e. "To what degree is it considered serious?",
- the necessity of government intervention in the issue,
- the global character of a policy plan,
- the measure itself: Is the behavior requested perceived as being able to solve the problem (effectiveness)? Is the solution perceived as being realistic? Is it justified to put that burden on the shoulders of the target group?

A process of acceptance rarely takes place within a short period. This is one of the reasons why efforts to gain acceptance of a policy after it is assessed by parliament run into difficulties, unless there is already a sufficient level of acceptance in most segments of the target group. If not, activities to improve the level of acceptance need to be started earlier, as part of the policy-making process.

This leads to a new function of communication, not directed to implementing a policy program effectively but to improving the policy-making process, and to harmonizing the content of policy more with existing community attitudes.

Government Policy as a Product of Communication

In this approach, government specialists use communication to explore social processes in defining and solving problems. The principal objective is no longer to change the attitudes or behavior of the citizens, but to change government. Affected groups should be represented and the government should be actively engaged in stimulating discussion amongst them. Government should bring these groups together to improve social problem-solving. To fulfill this role, a communication specialist must work according to four important principles:

1. The policy question must be flexible during the initial stages of policy development. If the problem definition and solutions are fixed, the public cannot become seriously involved. There must be scope for generating new results.
2. The information delivered by the government must be easily understood. This is difficult. Policy-makers are often incapable of talking about a problem in terms people can then express in their own words. They speak according to their own rationality, which can differ considerably from the rationality of ordinary people.
3. The policy process must be organized to be interactive, not just at a fixed point, but continuously, from the very beginning to the end of the process. Only in this way can people be involved and a sufficient level of understanding and commitment be attained. This flexible process leads to uncertainty, but it is only through interaction and understanding that the government can learn to think in terms of the rationality of the target groups.
4. A special effort must be made to shape a social learning process between groups in society. This approach is an answer to the failure in conflict resolution in government as well as in society. Traditionally, interactive policy-making has meant negotiating between government and special interest groups. The best result we can hope to achieve is compromise, where no one is really happy with the outcome. A better way is to involve both parties in a social learning process through which they gain an appreciation of why the other actor is talking in a certain way, and through which together they can seek new solutions.

Government as Communication

Building on the above, a completely new field for communication specialists can be identified. Rather than being concerned about government processes resulting in a fixed policy, they would be concerned with problem-solving without government intervention. The role of government is no longer to regulate, but to stimulate, facilitate or mediate. There is a growing conviction that government by talking and bringing people together is a more useful alternative than regulation, and relates very much to public network management. Based on this approach, a communicative infrastructure can be developed around certain subjects such as biodiversity and environmental problems in agriculture. Governments can subsidize conferences and journals. Communication specialists can become involved in the negotiation processes by mediating between different actors.

While some issues are being resolved in this way - e.g. agreements negotiated between the environmental movement and industry in relation to the packaging of goods - it is still too early to evaluate the outcome from a communication point of view. If social actors communicate because they realize that they are mutually dependent in striving towards their goals, they will learn about the rationality underlying the opinions of the other party. Shared decision-making can result from the ensuing mutual trust and respect. In this way, society fulfills its role without the expensive and often ill-considered regulation by government. The government's role becomes one of facilitation, bringing actors together, providing them with reliable information and stimulating communication among the general public.

However, a process of negotiation can turn into a situation where strategies for gaining power replace learning and consensus formation. Actors then construct negative pictures of each other and become actively engaged in spreading such images. Governments can take a non-cooperative approach by threatening to use forceful instruments such as regulations. This shift calls for a new role for government and communicators. The interaction within the government should be developed along three interconnected lines:

1 external to internal communication: bringing people in
2 internal communication: discussing how to adapt the content of policy to different interests and rationalities and
3 internal to external communication: how to inform people of existing ideas.

It is vital that communication specialists be integrated in policy-making processes, i.e. regular meetings with specialists, lawyers, technicians, financial experts etc. where they each take responsibility for the result. Acceptance is not based on outcome alone but on participation in the process itself. In this respect, two conceptions of the negotiation process can be differentiated:

- distributive negotiations where everyone is after a piece of the cake and
- integrative negotiations in which everyone is involved to create the cake they want.

Distributive negotiations start from fixed positions to which each party wishes to adhere as tightly as possible. People often ask too much, knowing that they will have to give something up. The negotiators are tight-lipped about their underlying motives and personal feelings. Threats are common and the constituency is kept alert with actively distributed images of the evil enemy. Sharing many lessons learned from alternative conflict management as outlined by Hamacher and Block in this PART, integrative negotiations proceed from an interest or an idea about the desired future. Understanding of the issues relating to biodiversity best comes from involvement in critical reflection. People are more open and try to share their feelings, beliefs and motives. Threats are minimized, and the functional relationship is kept as good as possible. Joint fact-finding is common. There is concern about the consequences of a discussion for the others. Most importantly, people learn empathy: the ability to see themselves with the eyes of the other, to be more socially reflective. Such learning processes do not exist in distributive negotiations. The more a negotiation follows the integrative model, the more successful it becomes. Interactive policy-making calls for the actors to learn from each other, to understand their interdependency and together elaborate more effective policy plans, the implementation of which is supported by different groups. However, the process is difficult and fragile, as attitudes stemming from previous situations are strongly embedded in social interactions.

Environmental Communication and Conflict Management

Winfried Hamacher and Karola Block

The authors of the preceding article on "government as communication" learned many lessons from conflicts between farmers and environmentalists, as pointed out in the Dutch de Peel case study in PART 5. Obviously there is a close link between environmental communication and conflict management which should be analyzed more here.

What is meant by conflict management?

Conflict management (CM) is designed as an alternative policy instrument, offering ways to build consensus and convergence in situations of open conflict and conflictive decision-making processes. A central premise of most CM procedures is that by separating the negotiation process from the disputed contents or issues, communication between the actors proceeds better and a commonly accepted solution may be found. So far, CM has mainly been applied to resolve disputes which are not part of a larger environmental communication (EnvCom) process, for example site-specific disputes about the location of landfill sites, or a policy dialogue on new conservation legislation. Nevertheless, CM can and should be used as an integral part of many EnvCom processes.

Why is conflict management needed in environmental communication ?

Although many appropriate tools and procedures are used in EnvCom, there are cases where communication approaches alone may not be sufficient. CM is frequently required for specific aspects of EnvCom, especially processes of social communication - promoting dialogue, reflection, participatory situation analysis, consensus-building, decision-making and action planning for change and development among people and institutions on different levels. CM offers appropriate approaches particularly when communication is deadlocked because the actors involved do not communicate in a constructive way, or some people are not reached by the EnvCom process and as a consequence the overall project cannot reach its goals. In order to move forward, the complementary use of other approaches and procedures becomes necessary.

When is conflict management needed within a communication process ?

The use of CM procedures within a communication process can be expedient in the following situations:

- The actors seem to have incompatible positions and interests.
- Communication is heavily burdened by prejudices, different values or political attitudes of the actors or by relationship problems between them.
- The previous EnvCom process was unable to clarify the actors' needs and interests, or may have even complicated the situation.
- Power imbalances obstruct open communication and consensus-building.
- The leading institution of the Env-Com process or the communicator in charge is not accepted or trusted by all relevant actors or does not have the capacity to deal with situations of conflict.
- Not all relevant actors can be involved in the communication process without a clear understanding of the dynamics of conflict and appropriate ways of dealing with them.

Training Workshop on Conflict Management in Indonesia

People from selected governmental and non-governmental organizations were invited to a general one-week training workshop on "Environmental Conflict Management". The objective was to promote CM. Due to the intensive collaboration in simulations and role plays, some of the existing "natural" barriers between organizations and different levels of hierarchy were leveled. This lead to new relationships nobody had envisaged in the beginning.

Additionally, CM may be an important part of educational communication. Especially educating young people to deal better with or to prevent difficult situations by early interventions which balance all actors' interests could be a valuable contribution to capacity-building for sustainable development.

In institutional communication, which aims to foster the flow of information necessary to improve coordination within the institution and between all actors involved in a development activity, CM does not provide additional benefit. This may change in those cases where, for instance, two ministries have diverging interests and no longer wish to cooperate. Very often, CM is used to solve conflicts between institutions which should cooperate to fulfill their tasks, enabling them to do so in a more constructive way.

How does mediation as one CM procedure work?

CM works with the decisive use of different procedures, such as mediation, conciliation or arbitration, benefiting from the specific advantages and areas of use particular to each. To fulfill the purposes of CM within EnvCom, the already widely known procedure of mediation may be appropriate. In mediation, a neutral third person - the mediator - primarily makes procedural suggestions as to how conflicting parties can voluntarily reach an acceptable agreement or consensus without a decision-making authority. Frequently the mediator works with the parties individually to explore acceptable settlement options or develop proposals that will move them closer to an agreement. With this procedural assistance, the parties in conflict are able to concentrate on the real issues/content and to negotiate solutions. Process-related factors which have so far burdened and prevented constructive communication can be clarified. Such factors can include a previous incident between actors, relationship problems between participants, financial constraints of the organizations involved or persons etc. The cultural and political background of the participants and the mediator considerably influences the practical form mediation takes. In any case, it is helpful if mediation builds on traditional conflict resolution mechanisms.

What has to be considered when CM is used within an EnvCom process?

A few concerns should be observed when CM is used within an EnvCom process:

- *At what stage in a communication process can and should CM be used?*
  Although the use of CM to resolve a difficult situation within a communication process is helpful at any time, early use increases the chances of obtaining a jointly supported solution in a short time period.
- *In whose interest is a CM process?*
  For CM procedures to be successful, it is important that all relevant actors be interested in resolving the conflict because the accomplishment of their interests depends on what the other actors do.

- *Is there a budget for CM?*
  As any EnvCom process, CM also needs to be considered at an early stage in planning the project budget.
- *Who could be the initiator and the mediator?*
  Any CM process needs an initiator - a person or an institution involved - who or which explores the need for and possibilities of CM, makes suggestions on the appropriate procedure and the relevant participants, finances it and - if necessary - looks for a mediator who should always be acceptable to all participants. At this point it is important to find out what role the communicator of the EnvCom process has now and had previously, and on which side he/she is perceived. In many EnvCom processes professional communicators have experience in and knowledge about CM and may even fulfill the role of a mediator, whereas in other cases it will be necessary to entrust professional staff with this task. In more traditional societies, traditional leaders or other elderly respected persons may very often be the best mediators, sometimes advised by a CM specialist.
- *What results are to be reached and how will they be integrated in the EnvCom process already designed?*
  While using CM within EnvCom processes, it is important to identify the issues participants wish to clarify and resolve in order to continue with the EnvCom process. In any case, the negotiated agreement should specify how the results will be implemented and monitored.

How could Conflict Management work within an Environmental Communication process?

Imagine a place on the coast of an African country where development institutions together with government and local authorities are supporting innovative approaches that are in line with the needs of local people while securing the sustainability of the coastal natural resources. More than 200,000 people in the region depend on fishing in the coral reefs as their main source of income. Destructive fishing practices such as dynamiting are a serious environmental problem and at the same time threaten the harvest from the sea, and thus the future of the communities. To address this situation, the institutions involved start with well-planned communication efforts, building partnerships between the coastal communities, the regional and district governments. Furthermore, communities are supported in developing their own action plans that control fishing, outlaw destructive fishing methods and close off parts of reefs to harvesting.

After a few meetings to develop an action plan in the pilot village, the difficulties of that approach emerge: the positions and interests of the relevant actors/participants in the process are very diverse, especially between the fishing families employing environmentally sound methods, those who use dynamite and the local government, which is represented by a technical specialist and a communicator in charge of this process. Additionally, there are two important and dominant persons who - due to former political incidents - try to argue permanently during the meetings, preventing a constructive atmosphere. The communicator is perceived as a representative of the local government and is powerless to help. Communication is deadlocked, and every meeting seems to make the situation worse.

In this context, the regional government together with the supporting development institution takes the initiative for a mediation process. They consider this helpful in identifying the real problem. Of course they would also like to reach an agreement, which seems to be a precondition for continuing with the whole project. In addition, this would set a good example and facilitate the process for the next villages.

As a first step, the regional government sends a professional CM specialist to propose mediation to the participants and to find a mediator who will be accepted by all participants. Initially, the people are not really convinced but the situation changes when an older, respected person from the neighboring village is chosen as the mediator. Coming from another village and not being directly involved, he is neutral enough to mediate and at the same time, he can integrate traditional conflict resolution schemes into this mediation process. During the next meeting the mediator and the CM professional jointly introduce and discuss the meetings and their goals with all participants. They also agree on some ground rules, facilitating the negotiations. In the following six meetings, agreement on the fishery practices combined with compensation measures is negotiated and accepted overall. Participants also decide that they can now develop the other parts of the action plan together with the local government communicator, knowing that if difficulties arise again, they can be helped by the former mediator. Additionally, they have all learned some new alternative and constructive ways to deal with difficult situations of dispute.

Tool - Negotiation and Mediation

Sometimes conflicts cannot be resolved so easily and more structured processes have to be used. Whichever approach is adopted, it must be appropriate for the context in which it occurs, and must take into account local customs and institutions for dealing with conflicts. There are three broad categories of approaches to managing conflicts. They differ in the extent to which the parties in conflict control the process and the outcome. These categories are:

Negotiation: where the parties, with or without the assistance of a facilitator, discuss their differences and attempt to reach a joint decision. The facilitator merely guides the process in a non-partisan manner to help the parties clarify and resolve their differences.

Mediation: where the parties agree to allow an independent, neutral third party (usually a person trained in mediation) to control and direct the process of clarifying positions, identifying interests and developing solutions agreeable to all. As with negotiation, this is a voluntary process which the parties can opt out of at any time.

Arbitration: where each side is required to present their case to an independent person who has legal authority to impose a solution. Agreements are enforceable through law.

General principles of negotiation/mediation.

To avoid focusing on particular stakeholders or positions, the best approach to adopt is what is sometimes termed "interest-based" or "principled" negotiation or mediation. This approach requires the parties to acknowledge that, to be sustainable, an agreement must meet as many of their mutual and complementary interests as possible. The focus should be on mutual cooperation rather than unwilling compromise. This approach encompasses four general principles which can be applied to conflicts in conservation initiatives almost anywhere:

Focus on underlying interests. "Interests" are people's fundamental needs and concerns. "Positions" are the proposals that they put forward to try to satisfy those interests. A conflict management effort in which all interests are considered is much more likely to result in a lasting and satisfactory resolution than one where the interests of only one side are addressed. Compromise may be the best way to serve everyone's interests in the long run, however, especially when overt conflict is replaced with the stability and predictability of a mutually agreeable solution. For example, in the context of the management of a protected area, allowing some use of the area's resources might ultimately serve the interests of conservation better than keeping the area in strict reserve status, and might also serve the interests of adjacent communities as well. The alternatives — which could include uncontrolled poaching or outright warfare — could be considerably more damaging.

Address both the procedural and substantive dimensions of the conflict. "Procedural" issues can include a group's need to be included in decision-making when their interests are at stake, to have their opinions heard and to be respected as a social entity. "Substantive" refers to interests that relate to tangible needs, such as availability of firewood, protection from predatory animals or protecting the land from damage caused by overdevelopment.

Include all significantly affected stakeholders in arriving at a solution. Failure to involve all affected stakeholders in the establishment and design of a conservation initiative, in decisions affecting management, or in working out how to resolve conflicts, generally leads to unsustainable "solutions" and to new conflicts arising in the future.

Understand the power that various stakeholders have, and take that into account in the process. Each party's approach to the conflict will depend on their view of the power they have in relation to the other stakeholders. For example, a group that feels powerless to influence an outcome through a bureaucratic process may resort to illegal activities instead. There are often extreme differences in power between the various stakeholders. People living in the vicinity of a conservation initiative may be poor and lack a formal education. Despite their lack of power, they should be included in reaching settlements to ensure that their needs can be met within the provisions made for the conservation of natural resources.

Conditions for negotiation and mediation

There are a variety of conditions which can affect the success of a negotiation. They should be present before a negotiation process is undertaken. The conditions are:

- All the people or groups who have a stake in the negotiations should
  be willing to participate.
- Parties should be ready to negotiate. They should be psychologically
  prepared to talk to each other; they should have adequate information; and an outline of the conflict management process should be
  prepared and agreed to. This is particularly important when dealing
  with different racial/ethnic groups, especially those which have a
  tribal system where speaking rights are subject to tradition and the
  consensus of other members. The negotiation/mediation process
  should allow time for the different cultural decision-making time
  frames to be accommodated, e.g. to select a spokesperson and to
  decide on the approach to be taken.
- Each party should have some means of influencing the attitudes and/
  or behavior of the other negotiators if they are to reach an agreement
  on issues over which they disagree.
- The parties should have some common issues and interests on which
  they are able to agree for progress to be made.
- The parties should be dependent on each other to have their needs
  met or interests satisfied. If one party can have their needs met
  without cooperating with others, there will be little incentive for them
  to negotiate.
- They should be willing to settle their disagreements. If maintaining
  the conflict is more useful to one or more parties (e.g. to mobilize
  public opinion in their favor) then negotiations are doomed to failure.
- The outcome of using other means to resolve the problem should be
  unpredictable. If one party is sure of complete victory for their point
  of view if they go to court, or directly to the government, they are
  unlikely to negotiate a settlement where only some of their interests
  will be met.
- All parties should feel some pressure or urgency to reach a decision.
  Urgency may come from time constraints or potentially negative or
  positive consequences if settlement is or is not reached.
- The issues should be negotiable. If negotiations appear to have only

win/lose settlement possibilities, so that one party's needs will not be met as a result of participation, the parties will be reluctant to enter into the process.

- Participants should have authority to actually make a decision.
- The parties should be willing to compromise even though this may not always be necessary. On some occasions, an agreement can be reached which meets the needs of all participants and does not require sacrifice on the part of any.
- The agreement should be feasible and the parties should be able to put it into action.
- Participants should have the interpersonal skills necessary for bargaining as well as the time and resources to engage fully in the process. Inadequate or unequal skills and resources among the parties may hinder settlement and should be addressed before negotiations commence.

Steps in the negotiation or mediation exercises

The process of negotiation can be viewed as comprising 13 basic steps. These steps can be used as a checklist for anyone called upon to facilitate such a process. The steps give no indication of the time required to complete them. The actual negotiation/mediation process may take a number of sessions. If the need for more information is identified at any point, the process should be stopped until that information is provided. If the parties reach a point where no progress is being made, they may decide to break off the process and either get back together at a later date or enter into an arbitration process instead.

The basic steps are as follows:

1. Prior to the parties' meeting, check that all or most of the conditions listed above are present. This will require meeting with the parties individually to clarify their attitudes and positions.
2. Set a time and place to meet that is agreeable to all parties.
3. At the beginning of the negotiation, ask each party to explain their position clearly: what they want and why. They should not be interrupted except for points of clarification.
4. After all parties have stated their case, identify where there are areas of agreement.

5. Identify any additional information that any of the parties need in order for them to be able to understand the claims made by other parties. If necessary, stop the process until they can be provided with that information.
6. Identify the areas of disagreement.
7. Agree on a common overall goal for the negotiations (e.g. the sustainable use of a resource and the maintenance of livelihood for a particular group or community).
8. Help the parties to compile a list of possible options to meet this goal.
9. List criteria against which each option should be measured (e.g. urgency of need, feasibility, economic returns).
10. Evaluate each option against these criteria.
11. Facilitate an agreement on one or more options that maximize mutual satisfaction among the parties.
12. Decide on the processes, responsibilities and time frames for any actions required to implement the agreement.
13. Write up any decisions reached and get the parties to sign their agreement.

adapted from: Beyond Fences, 1997

Mainstreaming the Environment through Environmental Education, Training and Communication - EETC

Ronny Adhikarya

Background - Agenda 21 and the Earth Summit

One of the important resolutions of the United Nations Conference on Environment and Development (UNCED) in Rio de Janeiro in 1992 was Agenda 21. Among other issues, it specified the need to increase environmental awareness, and undertake specific public education programs to bring about positive attitudes and appropriate behavior of various segments of the population towards sound and responsible environmental and natural resources management as well as sustainable development. Years have passed since the Rio Earth Summit, where most country leaders had agreed to take action in preserving the environment, for instance , through environmental communication, education, and training programs. Yet, very few of these activities have been implemented, especially for the rural population, despite the fact that many environmental factors directly and seriously affect agricultural productivity and the well-being of farm families.

Nevertheless, the last couple of years have witnessed an increasing number of decision- makers in the agricultural sector who are interested in mainstreaming environmental conservation and natural resource management issues and concerns into broad-based agricultural policy formulation, program planning, extension, education and training activities. In response to this trend and in view of the above-mentioned needs, the Food and Agriculture Organization (FAO) of the United Nations, which recently established a new Sustainable Development Department, has also intensified efforts in promoting environmental education and communication and training among its member countries. Through its Extension, Education and Communication Service (SDRE), FAO initiated field-level environmental education, training and communication activities in close collaboration with strategic partner institutions in a number of Asian countries.

One of the strategic approaches in disseminating and sharing environmental issues or concerns with the public is by "piggy-backing" relevant messages, through existing communication channels which have large and regular clientele, in an institutionalized and sustainable manner. In most developing countries, the majority of rural families have relied on agricultural field workers for information and advice on agricultural and rural development. However, in many countries, the agricultural extension service is often weak and ineffective, extension workers are not well-trained and have a narrow scope of expertise, mainly limited to agricultural production technologies. As many environmental issues are directly related to sustainable agricultural development, there is now a need, and a new opportunity, to revitalize and strengthen these services by mainstreaming environmental education and communication through agricultural extension training.

The process, strategies and results of incorporating participatory environmental education and communication activities into agricultural training programs in six countries in Asia, supported by FAO are described and analyzed below. The participatory methods employed in planning, implementing and evaluating environmental education, training and communication - EETC, and the institutional and professional networking strategies among eight institutions from six Asian countries - China, Bangladesh, Thailand, Indonesia, Philippines, and Malaysia - are explained. Concrete results from these two-year development cooperation activities are demonstrated.

Most importantly for the purpose of this Environmental Communication reader, the lessons learned from and best practices in planning, implementing, and managing the participatory and collaborative EETC activities in the Asian region are drawn and offered. The experiences can and should be utilized for further similar programs and replications by other interested institutions or agencies, in sectors other than agriculture and outside Asia.

Strategies, Process and Context

Learning from Below: Needs-Based, Demand-Driven, Diagnostic Planning

The integration and development of EETC through agricultural extension programs started as early as 1987 in Indonesia within the scope of an FAO Technical Cooperation Project requested by the State Minister's Office of Population and Environment, in collaboration with the Ministry of Agriculture. The evaluation results from this two-year project concluded, among other things, that expansion of project activities to other areas would require: (1) wider involvement of agricultural extension workers in an institutionalized manner, (2) integration or incorporation of environmental education and communication into regular, ongoing, relevant agricultural extension programs and (3) systematic and professional training of extension workers on relevant environmental education and communication issues, strategies, and methods, by proper and authorized agricultural training institutions. The Agency for Agricultural Education and Training (AAET), Indonesia, whose staff were involved in the above project, felt there was a critical need to include environmental education and communication in the agricultural training curriculum, and concluded that a systematic and institutionalized effort to develop its staff capabilities in undertaking cost-effective EETC activities would be given high priority.

Partnership with Strategic Training Institutions: a Key to Sustainability

In view of the above-mentioned conclusions, FAO/SDRE agreed to collaborate with AAET - an agricultural training institution which provides in-service training to all 33,000 Indonesian extension workers - in developing institutional and staff capabilities to master the process and methods of designing and utilizing well-planned and pretested environmental education and communication training curricula, modules and learning materials (see Rikhana in PART 5).

As will be discussed later, the training institutions participating in EETC activities in other countries are also those that provide training to most of the agricultural extension and other field outreach workers. Such institutions are the key or best agricultural training centers in their respective countries, and have the official mandate and major responsibility to carry out in-service training for agricultural personnel. Furthermore, these selected institutions perceive and maintain that their mission or mandate is compatible with, and supportive of, environmental education and communication goals. The incorporation of EETC into their existing training curricula and extension programs is expected to generate an added value, or produce synergy, which will facilitate EETC program sustainability and institutionalization.

Participatory Planning: A Knowledge Partnership Network on EETC

Based on the encouraging results and positive experiences of AAET's EETC activities in Indonesia, a participatory EETC program through agricultural training was designed and planned for interested institutions in the Asia region. A Regional EETC Planning Workshop in Malaysia in 1994 brought together participants from 10 countries - China, Bangladesh, Thailand, the Philippines, Indonesia, Malaysia, Tanzania, USA, Australia and Italy . These workshop participants represented public-sector training and extension agencies, NGOs, universities and international organizations, and came from multidisciplinary specializations including environmental science and management, agriculture, economics, public policy, education, training, communication, rural extension, etc. They discussed the needs, problems and strategies for incorporating EETC in agricultural extension and training. Based on the EETC experience in Indonesia, participants agreed on EETC's conceptual framework, strategy and methods, operational processes and implementation procedures. These standard operating procedures or guidelines included participatory mechanisms which will not only ensure relevance and client-orientation of related programs, but also provide for horizontal knowledge interchange and experience-sharing among institutions and planners within a country, region or globally. Eight institutions from six Asian countries decided to undertake EETC activities. which were initially supported by FAO/SDRE with small grants (ranging from $ 7,000 to $ 10,000). This knowledge partnership network of EETC resource persons has proved critical in providing participatory consultations, peer learning and reviews, and quality assurance and standards.

Franchising & Wholesaling of Training: an Approach to Institution Building

From Quantity to Quality

The chief strategy of this EETC program is the aim of "franchising" related activities to qualified and interested training institutions, instead of having FAO/SDRE and/or FAO projects conduct such activities in an ad hoc manner. The preoccupation with the "number game" in terms of the quantity of training participants should be shifted to developing partner institutions into high-quality training multiplier agencies. Instead of "retailing" such EETC courses by itself, FAO/SDRE's limited resources were used to develop training institutions' capacity and their staff capabilities in planning, designing and implementing cost-effective EETC activities on their own. The concept of "wholesaling" to training institutions is an approach which focuses on facilitating its sustainability and institutionalization, rather than to initiate a donor-driven, and external-funding dependent program. Furthermore, the "franchising" of the conceptual framework, methods, operational process and implementation procedures to interested training institutions, not only transfers EETC know-how, but also ensures quality and standards. More importantly, it also transfers institutional responsibility for advocating and further "wholesaling" and/or "retailing" of EETC "product lines" based on local needs and priority problems.

The EETC activities, therefore, emphasized two important aspects: (1) providing opportunities for knowledge-sharing and partnerships between institutions/ agencies and countries through regular workshops, technical consultations and review meetings, at the institutional, country, and regional/global levels, and (2) providing technical support for participatory staff training in the EETC process and methods, such as needs assessment, training module development and pretesting, multimedia materials packaging and production, training of trainers, training module utilization for training of extension workers, etc.

Emphasize Trainers' Mastery of the EETC Process and Methodology

In view of the above, EETC activities are geared towards ensuring that participating training institutions and agricultural trainers master the specific process and methodology of integrating or mainstreaming environmental education into agricultural training and/or extension programs, rather than focussing on specific technical environment themes . It is, therefore, critical that EETC activities be carried out by a multidisciplinary team, with environment specialists providing relevant technical subject-matter inputs.

Other strategies which are geared towards facilitating the "franchising" of EETC activities in given training institutions include: (1) initiating related needs assessment whose results are used for existing training curriculum reviews and reforms, (2) participatory EETC curriculum design and training module/materials development and testing, (3) training of master trainers in module utilization, (4) piloting "tryouts" for extension workers, and (5) obtaining contents validation, and EETC legitimization as an integral part of the existing training curriculum or extension program, by appropriate authorities.

Increase Stakeholdership among Key Trainers and Local Champions

The EETC process and method guidelines were developed with a view to increasing stakeholdership among the environmental education trainers, extension or outreach workers, local leaders and policy advocates concerned. Methods to solicit active participation and involvement of relevant persons and institutions who or which may be willing to "buy in" the EETC activities are of critical importance. Participatory training needs assessment and problem identification methods can facilitate the strategic planning process for designing a relevant and demand-driven and problem-solving environmental education and communication program. Participation of key stakeholders at all stages of the EETC planning, development, implementation and monitoring process is

86

critical for transferring technical and management skills and program responsibilities.

Encourage Local Ownership
of EETC Product Lines

An important component in the EETC process is a participatory curriculum development (PCD) activity through a series of training module "writeshops" and contents materials packaging workshops conducted by a team of local trainers and a pool of multidisciplinary resource persons. This participatory training curriculum development approach emphasizes the importance of having learner-centered, client-focussed and needs-based EETC modules or materials, instead of using expert-driven and top-down oriented training packages, or modules which may not be relevant to trainees' needs and the specific environmental education and communication learning objectives. While such a PCD approach may be more difficult to undertake, time-consuming, and perhaps rather expensive, it is more likely to be cost-effective, locally relevant, and properly utilized, as compared to the common - but often unsuccessful and unsustainable - practice of obliging local trainers to use imported packages of training materials developed by internationally known and highly qualified experts. Ownership of EETC product lines, such as training modules, training course offers, training reports, training announcement publicity, training marketing materials, etc. will need to be explicitly accorded to the local participating institutions and staff members who have been responsible for developing these products, and should be widely publicized.

Inducing Quality Assurance and Standards

There is a difference between "wholesaling" and "franchising" of training. In the context of EETC activities, the concept of "franchising" is more suitable as it emphasizes the need for training quality assurance and standards. Such a concern for quality is especially critical where participatory activities are to be carried out by many training institutions in various countries, which will determine and prioritize EETC contents based on their own training needs. To ensure high-quality of EETC is no easy task. However, it can be facilitated by proper planning, and applications of educational methodology tools such as training needs assessment, curriculum development, instructional design, learning methods selection, contents packaging, pretesting, training of trainers, evaluation, etc.

Develop and Use Participatory
EETC Process & Method Guidelines

During the 1994 EETC Planning Workshop in Kuala Lumpur, the participants developed generic EETC process and methodology guidelines which have now been used in planning and implementing related activities. The guidelines also provided step-by-step standard operational procedures, including suggestions for the development of environmental education and communication training modules (EETM). These guidelines are continuously revised and improved based on actual results and experiences of participating training institutions and comments of the EETC network members exchanged during the annual EETC regional workshops, e.g. the 1995 Bali workshop and the 1996 Beijing workshop.

Encourage Participatory Peer Review,
Progress Monitoring and Results Assessment

Institutions participating in the EETC program also determined specific objectives, measurable outcome indicators, planned activities, implementation time frame and resource requirements. FAO/SDRE provides networking opportunities for participating institutions and EETC network members. These include convening a regional workshop on a regular basis to facilitate participatory peer review, monitoring of progress and accomplishments, and to obtain suggestions for improvements. This participatory consultation and assessment among network members and practitioners using their first-hand experience constantly scrutinizes and improves EETC quality.

Create Constructive Competition and
Consultation Mechanisms

Another innovative feature of the EETC program is the utilization of "rewards and recognition" to training institutions which have conducted successful activities and outstanding network members by inviting them to national and regional conferences and workshops, appointing them as regional consultants or resource persons to assist EETC activities in another country. The above-mentioned participatory consultations through regional and national workshops are effective means for developing a constructive competitive spirit among participating training institutions to showcase their best performances and demonstrate positive results. Sharing and learning from such actual experience are powerful motivation stimuli and constitute an effective educational process which can inspire and lead network members to further improve EETC quality and standards.

Furthermore, a panel of EETC resource persons, who are highly qualified and experienced in relevant fields, such as environment, agriculture extension, education, training, communication, etc., has also been established to provide "on-demand" advisory, consultative and troubleshooting services. They have provided useful conceptual ideas and practical suggestions during regional and national workshops, as well as made independent, objective assessments and comments to participating training institutions for further improving their EETC activities.

Link with other Relevant and Complementary Network Activities

The EETC program and its network members are also linked to another participation-oriented program on Population Education through Agricultural Extension Training (PEDAEXTRA) executed by FAO/United Nations, and funded by the United Nations Population Fund (UNFPA). Through this PEDAEXTRA project, 12 institutions in 12 countries - Tanzania, Ghana, Malawi, Kenya, Burkina Faso, Morocco, Nepal, Indonesia, the Philippines, China, Ecuador and Chile - have collaborated since 1994 in integrating population education into agricultural extension and training programs. Cross-fertilization of ideas and experiences has been accomplished by inviting some EETC network members to PEDAEXTRA regional or national workshops, and vice versa.

Process Documentation to Facilitate EETC Replications

While summative evaluation or impact assessment is an important tool to demonstrate the degree of effectiveness of EETC activities, it may not be adequate for providing specific explanations and insights into the process and methodologies employed in undertaking such activities. It is thus very useful to carry out a process documentation which points out critical issues and decision-making requirements in undertaking EETC activities as well as their contextual background or circumstances. Through a chronological description and analyses of successful or less successful decision-making processes conducted during planning, implementation and management of activities, important lessons can be learned, and technical and management operation generalizations can be suggested, for future replications and expansion of similar activities. Finally, it should be noted that it is very important to prepare and commit adequate resources to complete the "last-mile" tasks in consolidating, summarizing and disseminating the process, methods, results and lessons learned from EETC activities in a user-friendly, attractive and captivating manner, especially aimed at relevant policy and decision-makers, for further improvement, expansion and replications of EETC activities worldwide.

EETC Activities - Status and Results

Building Strategic Alliances in Environmental Education and Communication

In all EETC activities conducted by eight participating institutions in six Asian countries, an analysis of the critical interplay among sustainable agricultural development (incl. food security issues), rapid population growth and environment deterioration issues was conducted . Taking into consideration the results of training needs assessment, an inventory-analysis of matching critical and relevant environmental education and communication messages - in most cases, including population education issues - and agricultural extension and training contents was undertaken. The purpose of this exercise was to identify suitable entry points for the integration of environmental concerns or issues into agricultural extension messages and training contents. Mainstreaming environmental education and communication into agricultural extension and training programs thus requires active collaboration between the institutions and/or line agencies concerned. Hence, an interagency and multisectoral policy advisory committee as well as a technical steering committee was established to guide and coordinate EETC activities.

In communicating environmental issues to the rural population, six participating institutions are utilizing agricultural training centers which train extension workers or rural outreach workers as the main channel for environmental education and communication. The main target beneficiaries of these EETC activities include: master trainers of the training institutions concerned, trainers of extension and outreach workers, field workers of the public extension service, and selected farmers/community leaders. Two other participating institutions have different target beneficiaries. In China, the Center for Integrated Agricultural Development (CIAD) uses the Central Agricultural Managerial Official College to "lobby" high-level agricultural policy and decision-makers by advocating the need for, and importance of, environmental management policy and education issues for the rural population. The International Institute of Rural Reconstruction (IIRR) in the Philippines is developing and conducting EETC activities for rural outreach workers of NGOs in the Philippines.

Within the scope of EETC activities, strategic alliances have been forged among relevant partner institutions, such as public-sector agricultural extension and training institutions, environmental management agencies, rural development training and adult education institutions, NGOs, community-based development agencies, and multimedia development and training centers. The accomplishments to date can be summarized as follows:

- **Environmental Education and Communication Training Modules** All participating institutions conducted an environmental education and communication training needs assessment whose results were used as strategic inputs for designing the EETC curriculum and developing Environmental Education and Communication Training Modules (EETM), ranging from 15 to 25 instruction hours.
- **Training of Trainers and Extension Workers** Training of master trainers in the utilization of EETM was completed in all participating training institutions, where at least two batches of master trainers had been trained. Each trained master trainer is expected to instruct at least another 10-12 trainers. Each of them was to train at least 2 batches of 20-25 extension workers as part of their institutions' routine and regular training programs by the end of 1996/1997.
- **EETC Institutionalization Process** EETMs have been reproduced for wide distribution to potential trainers and interested users in all participating institutions. In Indonesia, Bangladesh, China, the Philippines and Malaysia, the EETMs have been personally endorsed by the highest policy/decision-makers concerned and officially adopted as the required training module for in-service training of agricultural extension workers. In Indonesia, the EETM has been adapted and utilized by several NGOs and a local government institution, using their own resources. In Malaysia, Bangladesh and China, training of extension workers has been carried out without FAO/SDRE support. In the Philippines, additional funding from other donor agencies was obtained to support EETC activities of the UPLB consortium and IIRR. In Thailand, the Continuing Education Center (CEC) of the Asian Institute of Technology (AIT) is applying the EETC process and methods, and utilizing selected parts of the eight EETMs for its own environment-related training course.
- **Regional EETC Workshops** Three regional EETC workshops have been convened - Kuala Lumpur in 1994, Bali in 1995 and Beijing in 1996. During these workshops, new EETC network members were invited to join, including participants from Nepal, Ethiopia, Malawi, Kenya and Egypt.

The lessons learned from and best practices in planning, implementing and managing the participatory and collaborative EETC activities in the Asian region should be valuable for similar programs and replications by other interested institutions or agencies, in sectors other than agriculture and outside Asia (see PART 6).

PART 4 - Strategy

**10 Steps towards an Effective Environmental
Communication Strategy**
Manfred Oepen   95

10 Steps towards an Effective Environmental Communication Strategy

Manfred Oepen

The systematic use of communication is essential not only in project implementation but also to the improvement of policies and programs designed to promote participation in support of sustainable development. Communication is a two-way process in which a combination of "top-down" and "bottom-up" flows of information and experiences is required to analyze a given situation, determine the characteristics of strategic groups or the key problem to be tackled in order to arrive at the best mix of policy instruments. The most carefully conceived laws, economic incentives or technological solutions will not work before the people concerned have been informed, asked their opinion and, ultimately, gained "ownership" over the changes and interventions initiated to solve the problem at hand.

Isolated ad hoc initiatives that are not integrated into a comprehensive communication strategy may cause inflated expectations in rational appeals and the cognitive dimension of messages. This is why a project should define up front for what and for whom information is meant and how beneficiaries are supposed to pop pop them into communication and action. This is best achieved in a systematic and comprehensive EnvCom strategy which is always an integral part of a larger project or program and makes use of step-by-step strategic planning as part of a project cycle:

Stage 1   Assessment
1. Situation Analysis and Problem Identification
2. Audience and (KAP) Analyses
3. Communication Objectives

Stage 2   Planning
4. Communication Strategy Development
5. Participation of Strategic Groups
6. Media Selection and Mix

Stage 3   Production
7. Message Design
8. Media Pretesting and Production

      9   Media Performances and Field Implementation
    1o   Process Documentation and Monitoring
          and Evaluation

These 10 Steps will be outlined in this PART in greater detail. Its basic ideas are derived from various sources (Adhikarya/Posamentier 1987, SPAN 1993, Rice 1989), but mostly from the Strategic Extension Campaign (SEC) approach of FAO (Adhikarya 1994), and from the author's own experience in the field. Case studies from the SEC context such as a Pest Management Campaign in Thailand or a well-researched and documented Rat Control Campaign in Malaysia will serve as a kind of "red thread" throughout most of the steps. A brief version of this 10-step strategy is also available as a practical orientation brochure by GTZ and OECD (GTZ 1999, OECD 2000), both also accessible as pdf-files that can be downloaded from the respective web sites (www.oecd.org, www.gtz.de/pvi).

Strategic Extension Campaigns - The FAO Experience

A Strategic Extension Campaign (SEC) is "a strategically planned, problem-solving and participation-oriented extension program, conducted in a relatively short time period, aimed at increasing the awareness/knowledge level of identified target beneficiaries, and altering their attitudes and/or behavior towards favorable adoption of a given idea or technology, using specifically designed and pretested messages, and cost-effective multimedia materials to support its information, education/training, and communication intervention activities" (Adhikarya 1994: 4). What makes SEC special is

- the use of participatory Knowledge, Attitude and Practice surveys (KAP) to determine high-priority messages and the most strategic methods of communicating the message for maximum educational and cost-effective results,
- the participation of beneficiaries - not just "target groups" - and the social and media channels concerned with problem identification, planning and communicating extension messages.

SEC as an approach to improving the performance and impact of extension should not be applied indiscriminately. Hence, it may be considered and applied in countries or regions within a country where

- there is an organized extension service that has been functioning for a number of years,
- the extension organization has a minimum of trained staff in extension planning, training, and media planning and production,
- the extension service has facilities or potential access to facilities for media design and production,
- there are regular extension agents or workers who are trainable,
- there are "packaged solutions" available which will be the subject of the SEC such as low-cost composting, Integrated Pest Management - IPM, soil and water management etc.

Nine generic principles or aspects of SEC should be observed (Contado 1997):

- A strategic extension campaign has a well-defined objective, is problem-oriented, participation-oriented and focussed on a specific issue or recommended technology.
- Its goals are consistent with, and guided by, the overall sectoral development policies and extension program objectives.
- Campaign objectives are specific and based on the intended beneficiaries' felt needs and problems identified through a baseline survey of their Knowledge, Attitude and Practice (KAP) vis-à-vis the recommended solutions.
- A specific campaign strategy is developed with the aim of solving problems that caused non-adoption and/or inappropriate or discontinued practice of the recommended technology.
- A strategic planning approach is applied in the process of target audience segmentation, multimedia selection, message/information positioning and design, and extension/training materials packaging, development and production, with a view to obtaining optimum output/impact with the least or minimum efforts, time, and resources.
- Formative evaluation in the form of field pretesting of prototype multi-media campaign materials is conducted before they are mass-produced.
- Comprehensive and detailed campaign management planning is an integral and vital part of the SEC process. It will not only spell out the implementation procedures and requirements, but will also be used to develop a management information system, including monitoring and supervision procedures.
- Social briefing and training for all personnel who are involved in SEC activities must be undertaken to ensure that they understand their specific tasks and responsibilities and have the necessary skills and support materials to perform such tasks effectively.

Rat Control Campaign in Malaysia

As an illustration, the SEC in Malaysia on rat control in rice will be used because of its carefully monitored activities. The questions were: What do farmers know of "Matikus", a poison bait for rat control? The KAP data showed that 61 percent of the respondents were aware that Matikus was a rat control measure. After the SEC, 98 percent of the same respondents knew that Matikus was a rat control measure. The Malaysia case illustrates the increased cost-effectiveness of the extension program using the SEC method on rat control. For one rice crop season, the SEC operation, including the KAP survey cost US$ 140,000. The rice fields saved from rat damage were reported to be 477 hectares. The estimated amount of rice saved in one season was about 1,885 tons, which was valued at US$859,000. It may be noted that the impact of the farmers' knowledge, attitude and practice changes on rat control as a result of the SEC could be extended to several more seasons. One of the principles of SEC is to maximize the use and involvement of existing local human and institutional resources, which reduces costs and enhances the productivity of these available resources. Additional costs could be reasonably offset by the efficiency that is generated within the extension system and the higher productivity and gains of participating farmers. If the money spent on SEC is regarded as an investment, the measurement of its returns should be carefully monitored, as in the Malaysia SEC on rat control. In this case, the outlay of US$140,000 was more than offset by the value of 1,885 tons of rice saved, with an estimated value of US$ 859,000. One can stretch the analysis to include the impact of the US$140,000 SEC in the second and third cropping season.

- Process documentation and summative evaluation to assess the progress of implementation and impact of SEC activities are conducted. The results are used to improve ongoing performance, and to determine SEC's results and overall effectiveness as well as to draw lessons learned from such experience for future replications.

EnvCom Strategy Planning

The number of steps in an EnvCom strategy and, to a certain extent their sequence is not a fixed entity. One obviously cannot start with "M&E" or finish with a "situation analysis" but whether you define the problem or identify the audience first is often a matter of a team's individual preferences and style. For example, the renowned "Facts for Life" and "All for Health" family health campaigns of UNICEF, WHO, UNESCO and UNFPA use eight to twelve steps (UNICEF 1996:43); the EnvCom Planning Handbook of GTZ and METAP suggests six steps (Martin Mehers 1998); the Strategic Extension Campaigns of FAO employ ten phases (Adhikarya 1994). *Any* strategy should define and carefully plan the following elements: the objectives of both the project and the communication to support it, the strategic groups involved at various levels, the messages and media or communication channels to be used, the budget and the management. Also, it should be noted that the steps are not only interdependent but also represent an iterative process which moves in spirals, not in a straight line. Some of the EnvCom strategy's steps and elements are of a crosscutting nature, e.g. "participation" "planning" or "evaluation". Planners employ them at all stages of the process, not just at step 5 or 10 or at stage 2.

Planning is defined as a process of identifying or defining problems, formulating goals, thinking of ways to accomplish goals and measuring progress towards goal achievements. Planning must include strategy planning and management planning, i.e. the process of developing a strategic extension plan can be divided into two major parts. The first part is the process of *strategy development planning* ("what to do"), which usually comprises the first eight steps of the communication strategy as outlined above. The second part is the process of *management planning* ("how to make it happen"). When a plan for a strategy is completed, it must be pop popd into action. At that stage, the task of a communication planner shifts from strategy development to management planning.

It should also be noted that while an EnvCom strategy may and should incorporate a campaign, it is much more than that. In general, a campaign is limited to a relatively short period of time and it presents a readily available solution to a previously defined problem ("do things right"). A communication strategy starts before that, for instance with a social discovery process of questioning and researching a certain situation or policy in cooperation with the expected beneficiaries ("do the right things"). Also, it is not finished when the messages are disseminated through various media channels but it also takes responsibility for mobilizing and facilitating action that, ultimately, will lead to changes in the targeted environmentally harmful practices.

Situation Analysis and Problem Identification

There are many ways to conduct a situation analysis and problem identification. Participatory Rapid Appraisal (PRA) is one of the most participatory methods, which enables people to share, present and analyze facts that concern their life and development (see e.g. Chambers 1992, Schönhuth 1994, IIED 1995). PRA was adapted to environment-related and other methods such as Rapid Environmental Appraisal - REA, Participatory Urban Environmental Appraisal - PUEA, Community Self-survey - CSS, Social Impact Assessment - SIA and others. It can easily be combined with an analysis of Knowledge, Attitudes and Practices (KAP) of the actors or groups concerned (Step 2) and the formulation of situation-specific communication objectives (Step 3). PRA tools answer the "What? - Who? - Where? - When? - Trends?" - questions of any given situation analysis. For details see the tool boxes below.

In order to enhance the degree of participation and validity of PRA, it is recommended to hold a one- or two-week training event in which the staff of the implementing agencies, intermediaries (e.g. NGOs, media) and the stakeholders or actors concerned jointly participate. Once a mode of cooperation is established between those groups, they will interact and share experiences in other stages of the communication strategy as well, e.g. in pretesting media and messages, in utilizing traditional and community media as well as modern mass media, or in evaluating the success of activities.

PRA is structured by "triangles"

- teams - comprising men and women, old and young, multidisciplinary orientations, insiders and outsiders,
- sources of information - events and processes, people, places,
- tools and techniques - observations, diagrams, interviews and discussions.

The overruling principle of these triangles is participation from co-option and cooperation via consultation and collaboration to co-learning and col-

PRA
- is flexible and informal is applied in thecommunity by on-the-spot analysis
- works by optimal ignorance and appropriate imprecision
- avoids biases by being self-critical

lective action. PRA is processed in stages and by means of participatory tools:
• rural protocol • transect walk • mapping of observations • seasonal calendar
• problem-ranking by individuals and groups • pairing of problems related to
potential projects or interventions • data analysis • designing a development
plan • tackling constraints.

Although this reader does not provide sufficient space to outline the PRA method,
an example from a rural development project which used PRA in its community communication strategy is depicted briefly (Oepen 1996).

Participatory Rapid Appraisal

For two weeks, participants in an international training course practiced a community communication strategy in two villages (Candi and Kemusu) north of
Yogyakarta in Central Java, Indonesia. The training was embedded in an ongoing rural development project of PUSKAT, a local NGO, which also followed up
on the training with the villagers who actively joined in the PRA. The early PRA
steps took two days, for which participants were given a short guideline for
later exchange of initial experiences.

Day 1 and 2 in the Villages - Situation Analysis
"When you are doing your situation analysis in the village - e.g. using a transect
walk, mapping, focus group and household interviews - try to answer the following question to the plenary on Day 2: What are the main characteristics,
problems and needs of the village community you have interacted with in
terms of environment, demographics, infrastructure, health, social organization and culture, economy, gender and agriculture? Is there any difference
between your perception and the perception of the villagers? What conclusions do you draw?

Day 1
Rural protocol   The "rural protocol" upon arrival of the visitors in the two
communities varied considerably. In **Kemusu**, the international group was
welcomed by the local authorities - the head of the village, the division chiefs
of different local departments, farmers' associations etc. – in the village hall,
and was offered a formal platform to present their request to engage in a community communication field exercise with the villagers. The formal leaders
then gave their consent, offered the hall as a meeting place, and introduced
the visitors to individual farmer families and youth who were willing to cooperate.

In **Candi**, the community had been informed about the visitors but no formal welcome was prepared. Instead, the group took a stroll through the main street of the village and started talking to people outside their houses. The ones most interested and curious were the teenagers. They accompanied the group when it was explained to them who the visitors were and what they would like to do in the village. This way, more locals were introduced to the group and joined the walk. The house of a wealthy young farmer, who was also the head of the local radio listening club, became the meeting place.

Transect walk   During the transect walk, issues like land use and cultivation, resources and infrastructure available, state of sanitation, size of family, demographics, social organization and groups, indicators of economic activities, obvious developmental problems etc. were looked for, documented and discussed with the villagers who, in both cases, joined the participants in small groups. The various groups in each village coordinated their movements such that all parts of the village were covered in an irregular zigzag course. At this stage, individual or group interviews were only conducted informally and casually while passing through. At times, however, when a phenomenon of special interest came up which either the locals indicated or the visitors asked about, a group of walkers would stop in one place for discussion. The transect walk takes about three hours.

Mapping   The mapping of everything that was observed and noted in the village was done on the basis of visual sharing. Villagers themselves - in Candi it was mainly women - drew the map on a large scale (min. 5 x 5 meters) on the flat ground in public. They used sticks to draw and local materials like stones, leaves, plastic toys and bands, beans etc. They marked streets, rivers, water holes, buildings and other relevant items such as the mosque, cemetery, school, market, meeting places, water pipes, public toilets, refuse dumps, main farms, fruit-tree areas and other agricultural land etc. The villagers - young and old, women and men, elders and ordinary citizens - added to and changed the map according to the requests of fellow locals, or questions and information of the foreign visitors. While drawing all the items mentioned , the relationship between them and the meaning to the local people became clearer to the participants. From the villagers' perspective, the mapping gave them confidence in their abilities, established trust due to the respect paid to them and let them engage more deeply in a dialogue with the visitors, as well as in self-reflection about phenomena in their village. After the map was finished it was transferred onto a big sheet of manila paper by the participants so that it could be reproduced for their own purposes (see below). These map copies were stored at the village meeting place or brought back there for reference. Before the mapping, some teenagers had contacted the mosque and through its loud-speaker system, which is also used for village announcements, had asked more locals to come to the meeting place.

Information processing   Back to the meeting place, the participants discussed at a phenomenological level what they had observed. By then in Candi, more and more adults were coming forward, attracted by the call from the mosque . They joined the discussion on the findings of the day. In general, it was noted in both villages that

- people were predominantly making a living as "salak" fruit-tree farmers, with some paddy rice cultivation and home vegetable gardens left,
- income levels - indicated by the type of houses, walled-in compounds, number of TV antennae, cars and motor bikes – were relatively high
- there was little evidence of other trades or crafts, or of field laborers living within village limits
- in Candi, a number of huts and shacks on the village periphery were first taken for housing for such low-income groups but upon questioning, it turned out that these were sheds to separate the animals from the houses - another indicator for a relatively high development level
- the infrastructure and the maintenance of buildings and streets was well maintained
- sanitation and health facilities were well kept
- water supply by a pipe system from a nearby river and dug wells was regular, of high raw quality and regulated
- as the population was almost exclusively different Muslim, the mosque clearly dominated social life
- many households had access not only to the state-operated TV and radio stations but also to the private ones; some even owned video cassette recorders so that mass media exposure was high.

A first, necessarily rough picture of the two communities emerged. As a final activity, the groups decided in cooperation with the villagers which issues were still unclear or not yet covered so that they could be given priority in the investigations the next day. Then, more detailed and intensive research instruments such as interviews, family history or household visits would be applied.

Day 2

On the second day in the villages, the participants had a slow start with focus groups, household and individual interviews. This had to do with the work schedules of many community members. They suggested that the women and some of the teenagers would be available in the morning and afternoon while most adults would join after the afternoon prayer at the mosque. This arrangement was maintained for the rest of the "life case" so that, often, participants used the early morning for information processing and reflection, either individually or in a group, and stayed late in the village.

Interviews   The groups in each village agreed to use semi-structured interview guidelines and techniques and to coordinate in terms of topics and focus groups. A variety of different groups, informants households and random persons were interviewed in order to cross-check information, to make it detailed and specific, and to establish a personal relationship with a number of villagers. Questions were mixed with discussions while one group member always took notes unobtrusively. Leading questions and value judgments were avoided as the interview was intended as a respectful dialogue rather than an arrogant interrogation. "Why" questions, so far, were not much used for that reason. The groups used breaks for reflection on findings, discussion of the next steps and general observation. This whole process took about six hours.

Information processing   Back at the workshop venue in the late afternoon, the participants discussed the findings of the day again in the light of the previous day's questions and blanks . Some teenagers from Candi even followed the group to PUSKAT to learn more about the rationale of the international training. Also, they fully contributed to the discussions from their own point of view. Based on the discussions held on the preceding days, the findings and conclusions concerning the two relatively similar villages were elaborated in the two working groups:

- The farmers were all very poor until they changed from rice cultivation to "salak" fruit-trees in the mid- 1970's. Today, "salak" is very much in demand, increasing their income and dependency on that commodity.
- In both villages an unknown disease, a large insect, has infected many "salak" fruit-trees and threatened to lower productivity.
- The local agriculture extension service generally had a low credibility and was not experienced enough to be of assistance.
- Also, an expert team from Yogyakarta university that came and studied the situation could not offer any solutions.

104

- The farmers found that spraying the insects with pesticides or insecticides harmed the fruit-trees; they had not yet found appropriate means for biological pest management.
- A lot of people's relative wealth - relative in comparison to other rural areas in Java - was spent on the infrastructure of their homes, streets and fields, and on modern consumer goods like TV and VCR sets, not on productive investments.
- Many of the young were leading an affluent life that started corrupting their identity and negatively affecting their school performance and achievement motivation.
- Only the mass media were regarded as "entertainment", not the few, still existing local dance, music and song groups.
- A generation gap had developed as parents despised the leisure activities of their children - watching TV and video, riding around on their bikes - who showed little motivation to work the fields. On the other hand, parents realized they had "spoiled" their children.
- Some of the traditional art associations- like the "gamelan" orchestra – were still intact and active, but they were mostly frequented by the older generation, especially women

Key Problems   Next, a preliminary problem identification was carried out in the two working groups and presented in a plenary session as well. As the two communities were relatively homogenous concerning their general characteristics, their problems were also more or less the same:

Candi - a pest threatening the "salak" trees ? lack of local skills in appropriate pest management ? low credibility and availability of local extension service ? low learning and working motivation of youth ? social generation gap between parents and teenagers ? low entrepreneurial spirit ? "salak" direct marketing ? no senior high school ? no public telephone ? seasonal shortage of labor
Kemusu a pest threatening the "salak" trees ? inappropriate pest management through government-enforced pesticides ? low learning and working motivation of youth ? social generation gap between parents and teenagers ? low entrepreneurial spirit ? lack of self-reliance

Finally, the groups prepared the ground for the next step by discussing which issues should be focussed on in the data analysis the following day. Again, this was done in cooperation with the villagers themselves - a contact that was especially close with the Candi and Kemusu youth.

PRA, as indicated above, offers quite a range of methods and instruments which cannot all be presented here. Yet, at least a few tools should be evaluated more closely as they can be employed especially during the first few steps of a communication strategy.

Tool Box

Transect Walks and Diagrams

Transects are observational walks across an area or through a village. The walks help identify important aspects of the local environment (biological, physical and social) which may be discussed on the spot. They can also be used to verify, through direct observation and discussions with people met along the way, the information gathered by other means. At the end of the walk, the information collected can be summarized in a transect diagram which includes the key environmental features identified, an indication of relevant problems and resources, etc.

Purpose
There are two broad categories of transects: social and land-use. The former can provide information on housing density and types, infrastructure and amenities, cultural and economic activities, etc. The latter focuses on environmental and agricultural features such as cultivated land, forests, hill areas, types of soil and crops, and evidence of environmental degradation. The two can also be combined.

Steps in using the tool
- Decide which issues to focus on and what information needs to be collected .
- Identify local people to participate in the walk and explain to them the purpose of the exercise (three to five people will be enough to obtain a cross-section of views while keeping the discussions focussed).
- During the walk, take notes on relevant features. Seek clarification from people met along the way. Discuss problems and opportunities.
- After the walk discuss the notes with the participants and together prepare a transect diagram of the area covered. The notes and

diagram can be used in feedback meetings with the community at large.

Strengths and weaknesses
+transect walks are a highly participatory, simple and relaxed tool,
+they enhance the knowledge of local issues among all participants,
+they are useful for checking information shown on official maps,
+they can identify features not previously noted (because, for example, local informants assumed the researchers would know about them),
– they can be time-consuming,
– good transect diagrams require some graphic skills.

Trend Analysis

Trend analysis is used as part of an individual or group interview and consists of an in-depth inquiry on specific problems, how they have evolved, how they are likely to evolve in the future, and what actions need to be taken about them. For large areas, such as a region or country, trend-related data are often available, but for small areas, such as a village, it is unlikely that such data exists. Thus, the information required to show a pattern of change must be obtained locally.

Purpose
The purpose of trend analysis is to assess changes over time. Often it is used to raise the awareness of people about phenomena that accumulate rather slowly (e.g. soil degradation, population dynamics).

Steps in using this tool
- Decide what topic/subject you wish to assess.
- Help the community to decide on the accurate indicators of the subject. For instance, if the subject is community well-being, you could ask the participants what constitutes a good life for them. They may list household income, transport facilities, numbers of livestock, access to services such as education and health care, etc.
- Ask the participants to say where they think they are now in relation to each indicator, where they were 5-10-20 years ago and where

they think they will be in 5-10-20 years. Together with them, draw a graph of the trend for each indicator.

- To assess changes in the state of the environment or some specific species, you could ask the participants to list the main relevant plants or animals and then, on a horizontal axis, write the periods of time. Ask participants to either estimate numbers or the standard of well-being for each of the plants or animals at each of the points of time and record it graphically.
- Ask the participants to discuss the trends identified (e.g. What is happening? Why? Should something be done about it? What? What would be happening then?)

Strengths and weaknesses

+ creates an awareness of potentially negative and positive trends in the community, including the environmental impacts of activities,
+ group interaction enriches the quality and quantity of information provided,
+ different points of view existing in the community can be identified,
+ allows a comparison of trends in different indicators and, possibly, an estimate of the relationships between them,
+ cheap to use and can be adapted to the materials available (e.g. if there are no paper and pens, the graph can be drawn on the ground using leaves or stones as symbols and numbers);
- relies on memory and subjective judgments, although group interaction can control that to some extent,
- it is quite a complicated tool and needs the attention and very active participation of local people.

Focus Group Interviews

Focus group interviews are semi-structured discussions with a group of people (5-15) who share a common feature (e.g. women of reproductive age, shareholders in an irrigation system, users of a particular service). Participants are chosen by means of sampling procedures (e.g. from a cross-section of ages, a variety of land-area ownership, different resource users etc.). A list of open-ended questions is used to focus the discussion on the issues of concern but follow-up questions can be developed during the conversation.

Purpose

Focus group interviews were developed in market research to determine customer's preferences and expectations. Since the 1980s, they
have been used increasingly for sociological studies and in participatory
research, particularly research to identify and describe group perceptions, attitudes and needs.

Steps in using the tool

- Identify a list of key questions to guide the interview. Develop a
  system for analyzing the information collected, for example a matrix
  of topics and variables, or just a list of key topics and possible responses plus a space for comments.
- Identify the groups in the community concerned about the topic
  under investigation.
- Decide on the number of focus groups and the number of participants in each group. In a small community two or three groups (e.g.
  men/women, elders/adults/youth, agriculturists/herders, wealthy/
  poor) of five to ten participants each, may be sufficient.
- Conduct a practice (pilot) session with other community members to
  check that the questions are relevant and easily understood and that
  the type of responses can be summarized in the analysis system
  designed for the purpose.
- Before starting each focus group interview, explain the purpose of the
  exercise. Pose your questions to the group and be sure that each
  participant feels comfortable in speaking. Overtalkative participants
  should be controlled and quiet ones stimulated. Limit the duration of
  the session: A focus group interview should last about one hour.
- Since the interviewer also acts as a group facilitator, another person
  should record the discussion and jot down the essence of the contributions as well as the most characteristic quotes. If this is not possible, a tape recorder could be used, provided the group members
  give their prior permission. Tape recording is particularly helpful for
  reviewing the information in detail.
- Carefully review and analyze the interview notes or tapes to extract
  key statements, issues raised and patterns of responses in accordance
  with the analysis framework designed at the beginning of the process.
  The framework may need to be amended to accommodate unexpected responses.

- If possible, review the interview summary with the participants for them to check that their comments have been recorded and analyzed correctly.

Strengths and weaknesses
   + especially vulnerable participants may feel more free to talk when they are among peers,
   + group interaction enriches the quality and quantity of information provided,
   + different points of view between different groups in the community can be identified,
   - experience in qualitative research procedures is needed to use this tool effectively,
   - the facilitator needs to be able to stimulate group interaction during the interview,
   - the tool may entail interpretation of participants' responses by the person completing the analysis,
   - people may be reluctant to share their opinions with an outsider and some responses may not be entirely accurate.

Seasonal Calendars

Seasonal calendars are drawings or series of symbols illustrating the seasonal changes in various phenomena of an environmental nature (such as rainfall) or a social nature (such as labor demand or household income).

Purpose
The calendars generate information on seasonal variations in local problems, resources, constraints and opportunities. For instance, they can explore the use and reliance on various resources, the times when the community or specific groups are fully occupied, drought or flood seasons, hungry periods, cultural events, and so on. Calendars will differ depending on the occupations of the different stakeholders. For this reason it may be best to do this exercise separately with different interest groups.

Steps in using this tool

- Within a focus or community group, begin with a general discussion on the activities undertaken in the community throughout the year. This helps to focus the group on the task at hand. Make a list of all the issues or activities mentioned so they are not forgotten when the participants start to construct their calendar.
- Decide the appropriate format to use as calendars can be drawn in a variety of ways. The format and the symbols for the various items and activities should be selected by the participants. For illiterate people, symbols can be used to represent the months and activities. For instance, different lengths of sticks can be used to signify the different amounts of rainfall, or the availability of game in the forest. Another method is to draw a large circle with symbols representing the different months around the outside. The circle can then be divided into segments with symbols for different activities placed inside each of the segments.
- Once one or more calendars have been drawn, discuss the results. For information on labor demands, ask the group to estimate the proportion of time each spends on the various activities.

Strengths and weaknesses

- +seasonal calendars help the initiative staff to plan the best time to work with the community;
- +they help identify various local indicators for monitoring and topics for interview questions;
- +they illustrate the time variations in responsibilities and activities among different groups;
- – input can be very subjective and needs to be cross-checked by other tools, e.g. interviews with key informants or observational studies;
- – it may be difficult to estimate the seasonal changes in the various phenomena or the amount of time spent on activities — especially where the pattern changes throughout the year depending on product availability (e.g. water and fuel collection). To minimize this problem, ask for information in manageable time segments.

(adapted from: IUCN 1997)

Local knowledge systems

In recent years there has been a growing interest in integrating local knowledge into development planning and resource management systems. Unfortunately, this effort has largely involved collecting a limited amount of information about farming systems, forest management practices, and knowledge of traditional medicinal species. While PRA methods can help generate this type of information, current techniques often reveal only a small amount of such knowledge with respect to the system of resource management. What is perhaps far more fundamental is engaging local communities in meaningful dialogues and building on such knowledge. To do this, ongoing communication and a way for the sociocultural systems of disempowered groups, often linguistically different, to negotiate with modern, urban-based societies and governments is needed (Poffenberger 1997).

In the field of forest management, for example, much of the attention paid to local knowledge focusses on revealing the commercial potential of information regarding species utilization. Two inadequately studied but critically important categories of indigenous knowledge include local resource management institutions and land-use systems. Locally-instituted mechanisms to control access, participatory decision-making processes and conflict resolution procedures will be even more necessary to sustain the resources as population and economic pressures on forests increase. Understanding and supporting such institutions and their function will enable government agencies and other stakeholders to collaborate in the sustainably productive use of forests. Yet, planners are often from different social backgrounds, and even cultural and linguistic groups, than the communities for whom they are planning . Transferring information from indigenous, rural communities to urban-based organizations presents problems. A common frame of reference must be established whereby information can be shared and its implications assessed. Typically, local information is translated into the language of the urban planner and administrator, often losing meaning and specificity in the process (see IUCN 1997).

Complex systems of leadership, decision-making, dispute arbitration, and other components of local management systems also need to be identified in local terms - initially to facilitate communication between outsiders and community members, and in the long term to bring these elements effectively into collaborative decision-making. Local knowledge also extends to attitudes and beliefs. Of particular importance are local perceptions of resource rights, which are often based on the community's history in the area. Different communities

112

and other stakeholders may share a common history, but differ in their inter-pretation of past events and agreements. This information is important in de-veloping collaborative management mechanisms and agreements. Perhaps the first step necessary is mapping and documenting the existence of such systems. One Indonesian community forestry mapping specialist noted: "Just the pro-cess of recording the villagers' knowledge of their land and their history is empowering... The maps give the villagers some means of communicating with other land users, and some negotiating platform if there are conflicts." (Sirait, 1995).

Step 2

Audience and Knowledge, Attitude and Practice (KAP) - Analyses

The lessons learned from development communication and agricultural exten-sion teach us that if you ask people to change their practices - e.g. by recycling household waste or saving water - instructive information and raising aware-ness is not enough. The diffusion of an innovation requires (Rogers 1963)

Awareness	basic information about the new idea and how others use it,
Interest	the innovation to be applied to personal values and life-style,
Trial	attempts to practice the innovation and evaluate its usefulness and impact,
Adoption	acceptance and commitment to a change in practice.

Within a project life cycle of an innovation from awareness to adoption, com-municators distinguish early innovators (1o%), early majority (3o%), late ma-jority (4o%) and laggards (2o%). Communication strategists are well advised to identify the early innovators early on and integrate them as "agents of change" in addressing and convincing the majority.

Especially in EnvCom - where complex changes in attitudes and practices are at stake - this sequence is closely related to the potential barriers of communi-cation which were mentioned earlier in the "Said - Done" paraphrase: "Said is not Heard, Heard is not Understood, Understood is not Approved and Ap-proved is not Done". That is to say - if communicators cannot motivate and mobilize their audiences to take action and commit themselves to the new, environmentally friendly practices, raising awareness or creating interest indeed will not be enough. This process from *awareness* to *adoption* works best if the social groups concerned are actively involved and supported in a trustful partnership.

Social Actors and Stakeholders

While the broad term "social actors" is used here for the individuals, groups and institutions who interact with natural resources on any basis, the term "stakeholders" denotes those social actors who have a direct, significant and specific stake in a given territory or set of natural resources (Borrini-Feyerabend, Brown 1997). This may originate from geographical proximity, historical association, dependence for livelihood, institutional mandate, economic interest or a variety of other concerns. What is important, however, is that stakeholders

Community-based Groups or Organizations (CBG or CBO) are formal and informal groups of local people (e.g. a user group, a local cooperative, a village council, a residents' association) established to support the socioeconomic and environmental interests of their individual members or of the community as a whole. Their typical main assets:

- local knowledge, skills and resources,
- built-in flexibility,
- direct responsiveness to local interests and conditions,
- sociocultural cohesiveness with local communities,
- confidence and trust of the local people.

Example   The naam groups of Burkina Faso are traditional village youth associations composed of women (aged 15 to 21) and men (aged 20 to 35). The groups have several purposes, including promoting solidarity, cooperation, friendship and loyalty among the young and carrying out socially useful tasks. Typically, a naam group engages in paid activities such as harvesting for others or selling various products to collect money for a once-a-year festivity. As affiliates of a national association, the naam groups have begun channeling part of this money into development initiatives they run themselves, with impressive results.

(adapted from: IUCN 1997)

114

- are usually aware of their own interests in the management of the territory or set of resources,
- usually possess specific capacities (e.g. knowledge, skills) and comparative advantages (e.g. proximity, mandate) for such management, and
- are usually willing to invest specific resources such as time, money or political authority for such management.

Different stakeholders generally have different interests, different ways of perceiving problems and opportunities with respect to natural resources, and different approaches to conservation. They should all be equitably represented in developing an effective management system for the resources of common interest.

Non-governmental organizations (NGOs)

are non-profit groups acting in society on the basis of common concerns and specific capacities. Their typical main assets:

- professional expertise (knowledge, skills) in a specific subject (e.g. agroforestry),
- demonstrated effectiveness in pursuing common concerns,
- capacity to communicate and establish links at various levels,
- responsiveness and flexibility,
- social standing and autonomy.

Example   Zimtrust is registered in Zimbabwe as a welfare NGO; all full-time staff are Zimbabwean. The NGO aims to relieve poverty and improve the quality of life in rural areas. Its strategy emphasizes participation of local people in identifying, appraising, planning, implementing, monitoring and evaluating their own development initiatives. It also stresses the development of local institutions capable of managing natural resources while generating employment and income. The trust provides managerial, technical, material and financial support to several programs, as well as training in various skills. The best known program co-assisted by Zimtrust is the Communal Areas Wildlife Management Program (CAMPFIRE).

(adapted from: IUCN 1997)

The most basic stakeholders in the conservation of a given territory or set of natural resources are the people living within or close to them, usually grouped under the term "local community". In many situations these people are directly and highly dependent on the local resources for their livelihood, cultural identity and well-being. Communities are complex entities, within which differences of ethnic origin, class, caste, age, gender, profession and economic and social status can create profound differences in interests, capacities and willingness to invest in the management of local resources. What benefits one group and meets conservation objectives may harm another. For example, wildlife revenues may bring income to men, but more abundant wildlife may constitute a cost to women, e.g. because of crop damage. Even people sharing the same livelihood basis or personal characteristics such as farmers or unemployed youth should not be assumed to speak with one voice.

If it is rare for local residents to avoid diverging perspectives and conflicts, situations become even more complex when non-local stakeholders enter the picture. District administrators expecting to hold their post for just a couple of years, international conservation advocates, aggressive entrepreneurs, staff of national NGOs: they all bring forth particular views, capacities and interests. They both enrich and complicate the process and outcome of management.

When an agency aims at facilitating a management agreement among various parties, at least two crucial, difficult-to-answer questions need to be addressed. First and foremost: who are the "legitimate" stakeholders to take part in discussions and, possibly, in management roles? For instance, if a management agreement for a protected area has to be signed between a state agency and local residents, should parish-level representatives be involved or village-level representatives? To answer questions such as these, it may be useful to clarify and apply some considerations and criteria, which could include:

- existing rights to land or natural resources,
- continuity of relationship (e.g. residents versus visitors and tourists),
- unique knowledge and skills for the management of the resources at stake,
- losses and damage incurred in the management process,
- historical and cultural relations with the resources at stake,
- degree of economic and social reliance on such resources,
- degree of effort and interest in management,
- equity in the access to the resources and the distribution of benefits from their use,

- compatibility of the interests and activities of the stakeholders with national conservation and development policies, and
- present or potential impact of the stakeholders' activities on the resource base.

People's Associations
are district, regional or national bodies established with the explicit objective of representing the views and interests of a category of people (e.g. people with the same ethnicity, caste, gender, age-group, profession, etc.). Their typical main assets:

- large membership,
- capacity to serve the interests of the members,
- social standing and autonomy,
- accountability to members.

Example The Amazon's rubber tappers in Brazil are extremely poor; their economic survival is tied to the preservation of the tropical forest. Their union came to the world's attention because of its opposition to the clearing of large parts of the forests in the state of Acre to create cattle pastures. The union demanded that forest areas be designated by the state government as reserves where only non-timber products could be extracted. The interests at play were powerful and ruthless. Their leader, Chico Mendez, was assassinated in 1989 for his leading role in the union's struggle, but not in vain, since the following year the governor of Acre established four reserves according to the union's request. The largest of them covers almost a million hectares of forest, and has the support of the Brazilian government and of many international donors.

(adapted from: IUCN 1997)

On the basis of considerations such as these, it is possible to distinguish "secondary" and "primary" stakeholders, the latter sometimes also referred to as "strategic groups". This could then lead to different voices in decision-making and different roles, rights and responsibilities in management. It is important that the final number of stakeholders involved in management is well balanced: not too many so as to complicate and slow down the process and not so few as

117

to leave out key players. In all cases, it should be clear that people who believe themselves to be "stakeholders" are allowed to claim such a status and to argue their case on the basis of criteria such as those listed above.

Therefore, it is crucial to identify and analyze carefully

- the *stakeholders* and other *actors*, i.e. individuals, groups or institutions who have an interest or power relevant to the environmental problem in question,
- especially those later addressed as *beneficiaries* (or *strategic groups*), i.e. those addressed by the communication strategy and from whom a practice change is expected,
- and the *key intermediaries*, i.e. individuals, groups or institutions who can assist in reaching the target groups, often formal or opinion leaders, youth or women's organizations, NGOs who may lobby for public support etc.

Audience Segmentation

For the communication strategy as a whole, audience segmentation is very important. Relevant actors, beneficiaries and intermediaries are clustered into groups according to socioeconomic and other characteristics they have in common. In later stages, communication objectives, message appeals or participation options are analyzed and designed *per group*.

In audience segmentation, gender awareness plays a crucial role. Women's relationship to natural resources has evolved over generations and is embedded in culture. In more recent and "advanced" societies, women who carry the knowledge, experience and skills of utilizing and managing resources have been increasingly confined to a narrow household sphere. They have been deprived of traditional authority over resource management and have become dependent on market products. In addition, with the continuing fragmentation of land among family members, women's access to resources has shrunk. Over the last two centuries women's share of the formal economy and the production sectors, particularly in rural areas, has fallen dramatically. Agriculture replacing women's low-cost, self-sustaining management practices with short-term crop-specific external inputs has exhausted the resource base of the land. Even attempts to restore traditional and cultural resources and introduce "environmentally friendly technologies" assume a male-dominated institutional frame. "Organic farming" is an example. Women's long-standing relationship with natural resources is overlooked to the detriment of recovery (Wickramasinghe 1997).

118

Tool Box

Gender Analysis

In communities around the world, women as well as men are resource users and managers. Yet, in comparison with men, women tend to have different roles, responsibilities, opportunities and constraints, both within the household and in the community. An analysis of gender is therefore important to understand how resource users and managers relate to various resources and to each other. In some parts of the world, for example, laws and/or customs forbid women to own land, regardless of wealth or social class. This limits their options for independent resource management and land-use innovation. It can also lead to their losses being overlooked when compensation is provided for land acquisition for environmental initiatives.

Purpose
Gender analysis in an environmental initiative helps to illustrate the different ways men and women use natural resources, rely on them, and have access to alternatives. It also helps to make explicit the constraints (financial, legal, cultural, etc.) that affect the ability of men and women to respond to, and participate in, a conservation initiative. In this sense, stakeholder analyses, social impact assessments and evaluations should always include a gender dimension.

Steps in using the tool
Gender analysis can refer to any topic and be incorporated in all types of tools and processes, including:

- natural group interviews,
- gender-based interviews (natural group, focus and key informant),
- seasonal calendars,
- trend analysis,
- mapping exercises and
- household interviews (informal discussions)

Examples of questions for gender analysis (which can be asked of key informants, explored in gender-based focus groups or directly observed in the local community) include:

- Who has access to which resources (finance, equipment, land, natural products, etc.)?
- Who uses which natural resources and for what?
- Who carries out which tasks?
- What role do women play in decision-making about resource use?
- What is getting better for women/men?
- What is getting worse for women/men?
- Who is gaining from the conservation initiative?
- Who is worse off since the initiative began and why?
- Are there specific problems or constraints relevant to the initiative that apply specifically to women or men?

Women may be reluctant to attend meetings because of shyness or because the men in their families disapprove. In many communities it is necessary to employ female researchers to facilitate meetings with the women.

Strengths and weaknesses
+ ensures that the knowledge of women is made available in the design and management of a conservation initiative. This is particularly important in communities where the primary responsibility for agriculture and natural resource harvesting lies with women,
+ explicitly acknowledges the importance of the role and contributions of women in the environment;
+ protects women from having to bear unforeseen and unacknowledged costs which may result from the conservation initiative,
+ enables constraints on women's participation to be addressed, to facilitate their participation in the conservation initiative,
− patience and sensitivity are required of the initiative staff if women show reluctance to participate,
− addressing gender differences may be seen as a threat or criticism of the local culture and cause some resentment towards the management of the conservation initiative.

(adapted from: IUCN 1997)

Of course, audience segmentation is not limited to the differentiation of men and women. Other qualities such as age, education, income or class are equally important, often even within the same subgroup. For example, a pioneer waste separation project in Surabaya, Indonesia failed because the segmented audience "women" had not been differentiated into "housewives" and "housemaids". Even though the former had been persuaded to try separating waste, it was still the latter who had to do the job - but lacked information and training. Instruments and techniques that are useful to identify actors and relate them to each other include direct observation, interviews with individuals, focus group discussions or interviews, sociograms, resource users analysis. (see e.g. FAO 1994, Adhikarya 1987, IIED 1995, IUCN 1997).

Tool Box

Photo Language

Photo language is a way of using photographic images (pictures or slides) to promote reflection and awareness and/or collect specific information. Local people are trained to use a simple (or disposable) camera to take pictures of significant and good and bad features of their lives and their environment. It is important to recruit a variety of photographers (e.g. men and women, farmers and traders) as each will have a different perspective of what is relevant. The pictures or slides are exhibited and discussed in a group or community meeting.

Purpose
Photo language can be used for a variety of purposes such as participatory environmental assessment, gender analysis and appraisals of traditional and new technologies. Whatever their use, these tools entail an interactive approach. Slide language should not be confused with the use of pre-developed audiovisual materials for educational purposes.

Steps in using the tool
- Train several members of the community to use a camera and to compose and select significant images.
- Discuss with the group the purpose of the session and prepare with

them a list of relevant scenes to be photographed. Clarify with the group what each scene is meant to represent.

- If necessary, assist the group in taking pictures. Be sure that, for each image, several alternative shots are taken under different light conditions.
- After developing the photos, meet with the group and help them select the images they would like to show. Images should be relevant, easily recognizable by the audience and of good technical quality. A session usually requires 8-12 good photos or slides.
- Start the session by explaining its purpose, and then ask the people who took the images to describe and comment on them. For each image, have in mind a few questions to promote discussion if it proves necessary. If slides are used, project the slide long enough for the audience to identify the details and discuss the message. If pictures are used, they should be pinned up (if they are enlarged) or viewed around a table.
- Take notes on the main points of the discussion, possibly on a large flip chart or on a blackboard. Use them when wrapping up the session so that, before its conclusion, a list of the problems elicited from the images and possible solutions are considered.

Strengths and weaknesses

+ photo-appraisal and slide language are a creative and participatory way of identifying environmental/conservation issues and the various perspectives on these in the community concerned,
+ community members identify the messages and the scenes to be used and are encouraged to study and analyze their environment,
+ the combined images are likely to match the perspectives, priorities and values of the community as a whole,
+ it is an effective way of giving a voice to disadvantaged groups,
- slides are a relatively expensive tool. Cameras, slide film, a slide-projector and often a portable generator are required,
- photo/slide processing facilities are not always readily available,
- it may take some time for the tool to be properly effective as, at times, participants may be more attracted by the images per se than by the subject matter.

(adapted from: IUCN 1997)

Actors and Interests

When actors have been identified and segmented in relation with the environmental problem at hand, it is necessary to understand their interests because this will help to communicate with them more successfully. If a simple matrix of actors and their subgroups is not differentiated enough, the SWOT (Strengths, Weaknesses, Opportunities and Threats) is a useful technique to go into details (see e.g. GFA 1994).

The following chart from the context of an Indonesian Recycling Project (see PART 5) shows how the "Opportunities" and "Threats" - i.e. the future-oriented elements of the SWOT window which are based on the past related analysis of "Strengths" and "Weaknesses" - are fed into the communication strategy as benefits and costs of an intended change. "*Benefits*" are what is motivating, desirable, rewarding or pleasant about a practiced behavior or what the actors think they gain when changing their environmentally unfriendly practices. "*Costs*" or "*Price*" are what is difficult, unpleasant or undesirable about adopting a practice. The intended change, in this case, is the "4R" of solid waste management: "Reduce, Reuse, Recover, and Recycle Waste", which determine the selected entry points of greatest impact. Firstly, the SWOT of the new practice are analyzed for every strategic group. An opportunity (or benefit) for households, for example, may be additional income from recovered goods, while a threat (or price) to them may be the extra costs and efforts put into separating waste.

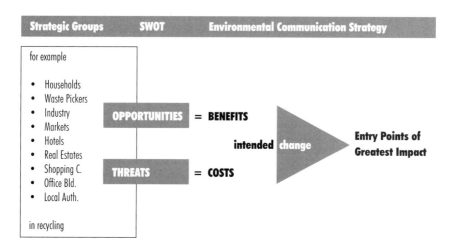

123

The selected entry point of greatest impact may, therefore, be to start with separating the more profitable and easy-to-handle goods such as paper, bottles or plastic and to link this separation of waste to the (informal) recycling sector.

Tool Box

SWOT Analysis

Strengths, weaknesses, opportunities and threats (SWOT) analysis is a structured brainstorming process to elicit group perceptions of a specific aspect of, for instance, a community, environment or project. The aspect is analyzed in terms of the positive factors (strengths), negative factors (weaknesses) at present, and possible improvements (opportunities) and constraints (threats) in the future.

Purpose
SWOT analysis can be useful for evaluating activities carried out in an environmental initiative. It can be focussed on specific aspects of the initiative, such as services provided by external agencies or activities being undertaken by a local community. It can also be used by stakeholders to clarify their views on a proposal before meeting with other interest groups.

Steps in using the tool
- A number of specific aspects to be evaluated are identified and listed one below the other on a board or flipchart paper.
- A four-column matrix is drawn on the side of the first column, and the four categories are explained to participants. To this end it may be helpful to phrase the four categories as questions e.g. "What are the good things about this particular service or activity, what has worked well?" (S), "What are the things that have not worked well?" (W), "What chances do we have to make things better?" (O) and "What things might work against us to stop us achieving the opportunities?" (T).
- For each aspect to be evaluated, listed in the first column, the group identifies the strengths, weaknesses, opportunities and threats, recorded in the relevant columns on the side.
- There are two ways to approach this exercise. You can go through all

the strengths and then all the weaknesses for all the aspects to be evaluated. Or you can go through the four categories for each item before moving onto the next item. Also, the time dimension can be used in a creative way: the OT of tomorrow will soon become the actual SW of "today", so that projections, expectations, fears or aspirations can be evaluated easily.

- Where there are different opinions about an issue, the facilitator should help the group to reach a consensus. Some points may need to be discussed at length. Comments are recorded in the matrix only after agreement has been reached.

Strengths and weaknesses
+ the technique allows positive and negative aspects to be identified and assessed for importance and therefore helps to set the basis for negotiations and trade-offs,
+ it can be a means of building consensus within a disparate group,
+ SWOT encourages group creativity and helps to link the perception of how things are with realistic expectations of how they could be, and to weigh the costs and benefits,
– a skilled facilitator is required for this process to be effective,
– sensitive subjects may arise which the facilitator may choose to return to later,
– conflicting opinions can be difficult to accommodate, which may make some people hostile to the process,
– some individuals may try to dominate the discussion,
– summarizing discussions into short statements requires a facilitator with good listening and interpreting skills.

(adapted from: IUCN 1997)

Hence, environmentalists in search of a key to human behavior and practices may find it useful to start with proven psychological determinants, such as:

Perceived benefits What advantage do people think they will get by adopting a new practice?

Perceived barriers What do people worry about, think they will have to give up, suffer, put up with or overcome in order to get the benefits?

| **Social norms** | Whom does the audience care about and trust on this topic, and what do they think that person/group wants them to do? |
| **Skills** | Is the audience able to perform the new action without embarrassment or without failing?" (Smith 1995: 11). |

Critical Behavior and Key Factors of Influence

In order to narrow down the field of practices potentially relevant to an observed environmental problem, communities in cooperation with communication specialists should consider

- the impact or importance of a particular behavior to the problem,
- the feasibility of changing or maintaining the behavior,
- whether the ideal behavior, or similar ones, already exist in the community concerned.

Critical Behavior

The practices which meet these criteria can be called critical behavior. In order to screen behavior that influences natural resources and environmental concerns it is useful

- to focus on specific forms of behavior rather than general categories,
- to emphasize the positive in existing practices,
- to classify behavior based on impacts it has on sustainability,
- to understand the feasibility of influencing relevant behavior,
- to understand behavioral flexibility.

RESOURCE	70's	80's	90's	REASONS
Crop yields	• • • • • •	• • •	• • • •	Pests Drought
Crickets	• •	• • • • •	• • • • • •	OUTBREAK ?
Water	•	•	•	Not enough Salty Far
Cattle Numbers	• • • • • •	• • •	• •	Drought Diseases Salty water
Goat Numbers	• • • • •	• • • • •	• • •	Salty water
Human Diseases	•	•	•	Water related
Construction Poles	• • • • • •	• •	• •	Population expansion
Available grazing	• • • • • •	• •	• •	Population expansion Fields increase
Population	• •	• • •	• • • • •	IMMIGRATION ?
Land availability	• • • •	• •	• •	Population expansion Fields increase

Historical Trend Matrix

Other tools that help achieve this task are (see e.g. Chambers 1991, IIED 1995, IUCN 1997 and tool boxes above)

- historical (trend) matrixes of (specific) resources and land use,
- ranking and prioritization techniques with respect to behavioral threats to sustainability,
- resource management decision charts,
- matrixes comparing the frequency of a specific behavior in various sub-groups within a community etc.

Key Factors

Understanding the key factors, motivational forces and influences related to critical behavior is the next step. In most cases, these include social, cultural, economic and ecological determinants. Potentially crucial ones are

Social factors	knowledge • values • social norms • cultural or religious values • skills • economics • laws • policies • gender etc.
Ecological factors	vegetative productivity • diversity • variability of physical environment (e.g. climate, seasons, daily periodicity) • history of disturbances • competition etc.

Tools that are useful in identifying such key factors among the many others that may be relevant to a given environmental problem are, in general (see e.g. IIED 1995, IUCN 1997 vol. 2, denkmodell n.d.),

- checklists of potentially important factors from: focus groups, community gathering, decision trees, pair-wise ranking, resource use trends etc.,
- techniques for identifying perceived benefits and prices: surveys, focus groups, comparisons of adopters and non-adopters,
- educational background, economic situation, gender, media access and other characteristics of the intended beneficiaries, cost-benefit comparisons etc.,
- causal webs and wiring diagrams: Venn diagram, social network maps, relationship wiring etc.,
- systems analysis: influence matrix, effects and axis diagram, force field analysis etc.

KAP Survey - Some General
Operational Recommendations

It is often assumed that "good" innovations sell themselves. The problem is, they do not. To start with, "good" is both a relative norm and a subjective term. It is obvious that the idea, product, technology or process constituting the innovation is of key importance to the adoption of the innovation. The innovation must be perceived to be desirable, an improvement over the present one, and worth the time or efforts - including social, opportunity, and financial cost - required for its adoption. The good quality and benefits alone, however, are a necessary but not a sufficient condition for the successful adoption of such an innovation. KAP surveys can provide data to help analyze the reasons for non-adoption of the innovation, and for developing strategies for overcoming such problems. It is often the case that non-adoption of a good innovation has nothing to do with the quality or the worthiness of the innovation itself. It was rejected due to the poor communication or extension process. Another major reason for the non-adoption of "good" innovations is related to, or caused by, non-technological factors, such as social, psychological, cultural, and economic problems or constraints.

Beneficiaries need to be consulted in the process of identifying problems and/or needs regarding their requirements or acceptability of a given innovation, i.e. a change in practice. A suggested procedure for conducting a participatory assessment of problems and needs is through a baseline survey on beneficiaries' Knowledge, Attitude, and Practice (KAP) on specific and critical behaviors and key factors. A KAP survey is oriented towards problem-solving and it operates at a microlevel, with a focus on determining at least three conceptual categories (see e.g. Adhikarya 1987, 1994):

- Knowledge, attitudes and practice (KAP) levels of audiences vis-à-vis the critical elements of a given recommended or intended innovation.
- The KAP survey seeks qualitative information from respondents, e.g. through focus group interviews, such as on the reasons for or causes of their negative attitudes and non-adoption or inappropriate practice with regard to the environmental problem.
- Information provided by KAP surveys is useful for campaign objectives or goals formulation and strategy development

KAP survey results can be utilized to analyze which specific elements of a recommended innovative, environmentally friendly practice or service are not known to the majority of beneficiaries involved, the reasons for their negative attitudes, how and why they carried out inappropriate practices previously etc. Therefore, in most instances, it is unnecessary to provide all beneficiaries with a complete range of messages and recommendations as some of them may already know, agree with or have acted on the necessary information. Strategic planning basically follows the principle of "start with what they know, and build on what they have".

However, without the benefits of a KAP survey that employs a rigorous scientific methodology for providing valid and reliable empirical data, the application of the above-mentioned strategic planning approach, and the objective evaluation of intervention impact cannot be done properly . Some important operational procedures in implementing a KAP survey include :

- review of the available innovative, environmentally friendly practice or service to be recommended,
- identification of essential elements of recommended practices, based on environmentalists' suggestions, and beneficiaries' inputs regarding their perceived priority problems and information needs,
- selection and prioritization of critical information which beneficiaries would need to acquire and master, in order to properly apply a given environmentally friendly practice,
- finalization of a critical information assessment to be used as a basis for developing a KAP survey questionnaire and focus group interview guideline,

- key informant interviews to validate and gather farmers' comments on messages regarding the recommended practices,
- focus group discussion to obtain qualitative inputs for the KAP survey, and to probe sociopsychological, cultural and economic aspects that may affect the non-adoption of recommended practices,
- development and pretesting of the data collection instruments,
- development of the study design and field implementation plan,
- implementation of field survey and focus group discussions,
- data processing, analysis and reporting,
- sharing and utilization of KAP survey results.

Normally, in an EnvCom strategy, two KAP surveys using the same data collection instruments need to be conducted, in order to provide empirical data for measuring the effectiveness of the communication intervention, and changes in the Knowledge, Attitude, and Practice (KAP) levels:

- the **Pre**-intervention KAP survey for baseline and strategic planning purposes,
- the **Post**-intervention KAP survey for summative evaluation purposes, often called Information Recall and Impact Survey (IRIS), and to be conducted at least 9-12 months after the first survey.

As a KAP survey is a participatory activity, most of the procedures are carried out in close cooperation with relevant participants, including researchers, subject-matter specialists, communication planners, trainers, extension workers, local leaders and citizens' representatives, as well as behavioral specialists and social researchers.

ABC - Six Steps to Applied Behavioral Change

In a nutshell, the various steps in situation, actor and KAP analyses can also be summarized in an Applied Behavioral Change model, which is often used in the context of social marketing approaches that can be integrated in the EnvCom strategy. The most crucial steps are outlined below and exemplified in the ABC case study in PART 5.

Social Marketing

The 10 Steps of a Environmental Communication Strategy or the six steps of the ABC Model are often combined with the key elements of social marketing approaches which have proven effective in family planning, health care and other fields where, just as with environmental issues, sensitive and complex behavioral changes are at stake. There are abundant social marketing case studies from both industrialized and Third World countries (see for example Atkin/Meischke 1989) This is why social marketing should increasingly be applied in EnvCom strategies.

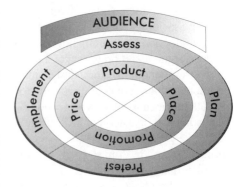

The field of social marketing did not enter the arena of communication science until the early 1970s, and was geared towards questions such as "Why don't people do what's Social Marketing

130

A B C - Model

Six Steps to Applied Behavioral Change

1 - Observe Behavior Identify what people like and don't like about a certain behavior that is to be changed. Don't just ask questions. Look, count, record behavior. Arrange for a few people to do what you would like the whole community to do. Watch their problems.

2 - Listen to People Ask what matters to them, talk about how your target behavior fits into their daily life. Look for what they get out of behavior as 'gain' or benefit and who matters to them.

3 - Decide What Matters Compare people who show the desired behavior with people who don't. What are they like, where do they live, how do they act out the behavior you care about? Segmentize your audiences because they will have to be communicated with differently.

4 - Generalize Facts Summarize critical environmental practices, key factors influencing behavior and other points such as benefits people care about, messages preferred, opinion leaders people trust. Test your assumptions with a representative survey.

5 - Deliver Benefits Deliver benefits people want, not just information. Solve barriers the people face, don't just 'educate' them. This means that service delivery and communication inputs have to be synchronized.

6 - Monitor Effects Find and fix mistakes. Selectively monitor crucial program elements by means of simple and manageable indicators for the behavior you wish to change.

good for them? Why don't they stop smoking, eat right, stick to one faithful sexual partner, or at least use a condom when they don't? Why don't people get their kids immunized and why do teenagers keep trying dangerous drugs and ruining their lives?" (Smith 1991). The search for answers in changing people's behavior has led to a simple definition of the new field. "It is the design, implementation and control of programs aimed at increasing the acceptability of a social idea or practice in one or more groups of target adopters." This definition emphasizes several important points about social marketing. Firstly, it is a program design and implementation process. It is not a theory or an idea - it is a process for organizing interventions. Secondly, social marketing works for ideas like "stop cutting trees" or ,"don't waste water" as well as it does for practices and products like environmental technology, composting methods or Environmental Impact Assessment. Finally, social marketing focuses not on everybody, but on target adopters - a specific subset of a population chosen because its members share a particular problem and important values that make it possible to address appeals to them that work for the entire subgroup.

The Concept of Exchange. Behind marketing is the notion that people do new things or give up old things "in exchange for" benefits they hope to receive. Unless we identify a benefit people actually want - a benefit we can offer "in exchange for" what we want from them, marketing argues that people are not likely to take our advice. The reality is that the "benefits" people want often have little or nothing to do with the ecological benefits environmentalists want them to care about, especially if the benefit is as abstract as "biodiversity", for example. To use a well-established example from the health sector (Rice 1989), teenagers are supposed to drive without drinking because the "benefit" they should experience is avoiding death. However, we know that many teenagers see death only as a remote possibility. Avoiding death is not perceived by them as a benefit, no matter how dramatic the TV spot on death, drinking and driving. But some teenagers do fear disfigurement - being scarred for life. "To avoid disfigurement" is a benefit some teenagers can believe in. The search for what to give people "in exchange for" what we want is at the very heart of successful social marketing.

The Marketing Mix. Social marketing is also practical. It offers four basic categories of inputs that program managers can organize and shape to give people what they want. Marketers, social or commercial, call these categories "The Four Ps". Shaping the four Ps is called developing the marketing mix. The first thing managers can shape is the product "P." For example, they can make condoms more attractive by adding lubricants or making them stronger, thinner, colored, ribbed or smooth. They can also shape an immunization service by making it more convenient, more fun and less time-consuming. Price is the second "P" managers can shape. This is a complex cluster when talking about social ideas but one which leads us to discover the real barriers people experience in trying a new health behavior. The manager's job is to reduce the cost or barriers people face. Place is the third "P". It includes the distribution system and sales force. Most environmental organizations do not usually think of managers, clerks, family members, or friends as a sales force - but they are. And marketing gives managers a powerful set of tools to help environmental professionals, peers and the family become more understanding, more articulate, better listeners and more sensitive to their audience's needs. Finally, there's the promotion "P", the TV spots, slogans, celebrity spokespeople, etc. This is an important program element but it is only one element. Promotion is the voice of the program which deals with messages and channels: what to say and how to get it heard. The most important insight of social marketing is the interrelations of the four "Ps", the balancing act that program managers must go through to choose the most important of the four inputs to shape for the product and the audience - which varies from one problem and one audience to another.

Audience Research. The intermingling of action and research is absolutely fundamental to successful marketing. The audience is the center of this intermingling. Marketing goes along not only with an appreciation for audience research but numerous tools - from intercept surveys, to focus groups and ethnographies that bring qualitative and quantitative techniques together to build a better picture of the audience, their needs and the benefits they desire. Few other program design systems place more emphasis on practical field research to guide program decision than social marketing. It is also a practical way to integrate service delivery with consumer demand. People must want to use new services. They must know how to use those services easily and then must be able to seek them out effectively.

Marketing is not the only strategy to change behavior. There are many strategies for social change: technological, economic, political, legal, educational. Smoking, a well-studied example used in this reader before, can easily illustrate this point. Technology could solve the problem of smoking. If we had a cigarette that did not harm anybody, smoking would no longer be an issue. There are economic behavior-change measures. If people could not afford to buy cigarettes, smoking rates would go down, as in California when 25-cent increases were mandated in the late 1980s. There are political and legal strategies. As it becomes harder to find a place to smoke, people smoke less. Education can also affect smoking behavior – with school-based programs and physician intervention programs being good examples.

Step 3

Communication Objectives

Communication objectives should be very specific and aimed at increasing knowledge, influencing attitudes, and changing practices of intended beneficiaries with regard to a particular action. A communication objective describes an intended result of the environmental communication activity rather than the process of communication itself. Once the problems have been identified and the stakeholders analyzed, the communication objectives should be defined. It should be pointed out, however, that communication objectives are usually *not* the same as the project or program goals, which are expected to be the ultimate results of the whole communication strategy *plus* other supporting outputs. The achievement of the communication objectives is a necessary, but not a sufficient condition for achieving the project or program goals. Hence, communication objectives should reflect the environmental policy, project or program goals, respond to the needs of the program and its target audience and help in solving the problems encountered in achieving such goals.

Communication objectives should specify some important elements or characteristics of the policy, project or program activities which could help to provide a clear operational direction, and facilitate a meaningful evaluation. Some of those elements are:

Inadequate
"To drill 4,000 ring wells and 2,000 tube wells by August 1994".

Comprehensive
"To increase the number of small farmers in districts X, Y and Z using water from the wells to irrigate their farmland from the present 100,000 to 175,000 small farmers within two years".

- the target beneficiaries and their location,
- the outcome or behavior to be observed or measured,
- the type and amount/percentage of change from a given baseline figure expected from the beneficiaries,
- the time frame.

134

Any policy, project or program goal should be explicit in specifying what is to be accomplished, not just the general or operational elements to be achieved. The descriptions of both project and communication objectives should be made more comprehensive and specific, and reflect the actual scope of the program. In the case of an irrigation program, examples of communication objectives which would support the achievement of general extension program goals could be:

- to inform at least 65 percent of the small farmers in X, Y and Z districts about the procedures and benefits of an irrigation system using ring and tube wells within one year,
- to reduce the proportion of small farmers in districts X, Y and Z who have misunderstandings and misconceptions about the cost and technical requirements of drilling and building ring or tube wells, from the present 54 to 20 percent in one year,
- to increase the proportion of small farmers in districts X, Y and Z who have positive attitudes towards the practical and simple use of the irrigation system to water their farmland, from the present 32 to 50 percent within two years,
- to persuade small farmers in districts X, Y and Z to use water from the wells to irrigate their farmland, and to increase this practice from the present 20 to 35 percent in two years.

In the example below, defining clear-cut campaign objectives for a "Pest Management" project in Thailand links the previous KAP survey to later stages of the campaign strategy, e.g. message design(see e.g. FAO 1994).

Step 4

Communication Strategy Development

At step 4, enough baseline data on problems, needs, actors, project and communication objectives is available to put all information in a context. The effectiveness of an environmental communication strategy depends very much on its planning, which should be specific and systematic. Planning is defined as a process of identifying or defining problems, formulating objectives or goals, thinking of ways to accomplish goals and measuring progress towards goal achievements. Strategic planning reflects the beneficiaries' identified problems and needs and the way information, education, training and communication

Specific and Measurable Campaign Objectives

Based on the Problems Identified by the KAP Survey for the

Strategic Extension Campaign (SEC) on Pest Surveillance System in Chainat Province, Thailand

	IDENTIFIED PROBLEMS	EXTENSION CAMPAIGN OBJECTIVES
1	Low knowledge on pest identification and nec-essary action for pest control	To increase the percentage of farmers who have knowledge regarding: a. Pest identification from 41% to 65% and, b. Necessary action for pest control from 15.1% to 40%
2	Lack of sufficient knowledge on the importance and benefits of natural enemies	To increase the percentage of farmers who know the identity of natural enemies (good bugs) from 11.4% to 35%
3	Lack of sufficient knowledge on the importance and benefits of resistant rice varieties	To increase the percentage of farmers who have knowledge regarding the recognition and importance of resistant rice varieties from 35.8% to 50%
4	Lack of awareness on Surveillance and Early Warning System (SEWS) programme and Pest Surveillance (PS) form	To create awareness by increasing the percentage of farmers having knowledge on SEWS from 13.2% to 50% and to increase the percentage of farmers skilled in the use of Pest Surveillance (PS) form from 10.1% to 30%
5	Farmers prefer broad-spectrum pesticides and blanket spraying	To reduce the percentage of farmers using broad-spectrum pesticides by: a. Increasing the percentage of farmers who know how to choose right chemicals from 5% to 16% b. Decreasing the percentage of farmers who prefer broad-spectrum pesticides from 65% to 50%
6	Farmers do not believe in the effectiveness of natural enemies	To reduce the percentage of farmers who do not believe that conservation of natural enemies can suppress pest population from 36.5% to 25%
7	Farmers go to the edge of the field, but NOT into the field to check for pests according to the recommended precedure and frequency	To increase the percentage of farmers who check their fields according to the recommended procedure from 17% to 35%
8	Farmers spray pesticides on sight of pests based on their "natural instinct"	To reduce the percentage of farmers who believe in the need for spraying pesticides as soon as pests are observed in the field, without checking the field properly, from 69.8% to 55%
9	Farmers are aware of pesticide hazards, but DO NOT apply safety precautions in pesticide handling, application and disposal	To increase the number of farmers observing adequate safety measures in using pesticides by increasing the percentage of farmers practising correct disposal of left-over pesticide from 10.7% to 25%

will be used in solving such problems or meeting the needs. Such a plan must outline the management actions to be taken in implementing the strategy. Strategic planning can be operationally defined simply as the best possible use of available and/or limited resources, i.e. time, funds and staff, to achieve the greatest returns or payoff, i.e., outcome, results or impact.

The process of developing a strategic extension plan can be divided into two major parts. The first part is the process of *strategy development planning* ("What to do") which comprises the first eight steps of the communication strategy as outlined above, i.e. up to message design, media pretesting and production. The second part is the process of *management planning* ("How to make it happen"). When a plan for a strategy is completed, it must be pop popd into action. At that stage, the task of a communication planner shifts from strategy development to management planning. Even though these steps will not be *implemented* until later, they need to be *planned* at this stage. To transform strategies into activities, management objectives must be identified clearly to include at least the following elements:

- what the action is,
- who is to carry out the action,
- how the action is to be carried out,
- the volume of resources needed and how
  to obtain such resources,
- when the action is to be accomplished,
- how to set standards for measuring progress
  and impact of implementation.

In addition to media performance and field implementation (step 9) and process documentation as well as Monitoring and Evaluation (step 1o), other management tasks include developing an exit strategy for the time after the program or project to which the communication strategy is related has been finalized, and identifying and meet training and skills needs of both field personnel and beneficiaries. There are at least three kinds of management activities for which regularly updated information is needed to make effective decisions - personnel, finance and logistics.

The chart below provides an orientation guideline on how to determine the general communication strategy direction and priority on the basis of KAP survey findings. This general strategy must be made more specific in steps 4, 5 and 6 of the planning process. The guidelines should not be used as a recipe

If:

Then:

Situation	Position of people involved concerning			Priorities of an Environmental Education and Communication Strategy			Fields of action and communication channels for environmental education and communication		
	K knowledge	A attitude	P practice	Main Approach	Main Objective	Didactical Emphasis	Mass Media Sensitiz.	Group Media NGO	Interpers. Commun. Consult.
1	low to medium	low	low	informative	awareness creation, increase of operational knowledge, identify needs and advantages	What + Why	high	low	low
2	medium	medium	low	informative motivatingd	identify needs and advantages, Inform about and demonstrate alternatives	Why	high	medium	low
3	medium	medium	medium	motivating action oriented	alternative problem view, discuss solution approaches, explore roots and consequences of negative activities, try out feasibility of solution proposals participatorily	Why + How	medium	high	medium
4	high	medium	medium	motivating action oriented	explore negative roots of atitudes, skills training through „learning by doing" for behavior change, correct counter-productive practices	Why + How	low	medium	high
5	high	high	low medium	action oriented	skills training through „learning by doing" for behavior change, logistical assistance and consulting, explore dissident attitudes and tackle roots	How	low	high	high

but as a tool to conceptualize and systematize communication strategy planning and development.

Examples for communication strategy development related to a "Pest Management" and a "Rat Control" campaign are presented in "Strategic Extension Campaigns" by FAO and other publications (see FAO 1994, Adhikarya 1987).

STEP 5

Participation of Strategic Groups

Participation is a process of motivating and mobilizing people to use their human and material resources in order to take their lives and their hopes in their own hands. The participation of strategic groups is such a crucial element in the Envcom strategy because people will not change their environmentally relevant practices if they do not have a say in planning, implementing and evaluating the action for change. That is why it should be considered as an individual step in the mainstream of the entire process. However, like planning or evaluation, participation should be a continuous, not a one-shot effort. The keyword here is ownership. It should be taken literally in terms of media products and communication processes not *for* or *about* people but *with* and *by* the people themselves. This procedure safeguards project or program sustainability and achieves the media mix that is best suited to the sociocultural circumstances. It is difficult to "own" TV, video, or radio because of the financial, technical and skills levels involved. It is much easier to "own" a people's theater production or other community media that are managed and produced by local means and towards local ends. This does not imply, however, that participation should be constrained to the "community media". Instead, strategic alliances with the "mass media" should be built that strengthen the "upward compatibility" of the communication processes - e.g. a local theater performance on people's action related to an environmental problem that is recorded on video, edited professionally and broadcast on TV as a feature film or newscast.

Participation incorporates all project levels - assessment, planning, implementation and M&E. Crucial questions are:

- Who sets the agenda on the general problems to be studied?
- Who says which needs should be met?

- Who is consulted in planning for appropriate solutions?
- Who determines which media will be used?
- Who carries out the action?
- Who produces the media and designs the messages?
- Who sets the standards for measuring progress and impact?

Benefits of Participation

To begin with, participation is a condition by which local knowledge, skills and resources can be mobilized and fully employed (Borrini-Feyerabend 1997a). Local people may understand very well the causes of and possible remedies for deforestation or soil erosion in their environment. They may know how to find and use plants of unique properties or how to prevent animals from damaging new seedlings. They may be able to offer labor, land, food, shelter or tools to run a project. Contributions like these increase the flexibility of an initiative and its responsiveness to local conditions. They also reduce the chance of mistakes with major environmental consequences, and often mean the difference between success and failure. In fact, the overriding benefit of people's participation is the increased effectiveness of any initiative.

Another major benefit is a more efficient use of resources. Local knowledge and skills help minimize waste and obtain results with limited investments. Participation can bring to the project the full benefits of human and material resources that would otherwise remain idle or poorly utilized, and local monitoring discourages the undue use of assets and promotes accountability and respect for rules. Most of all, however, the participation of local people provides a unique assurance of the sustainability of a conservation initiative. Usually, local people are those most directly interested in the positive results of such initiatives. In fact, most local communities possess greater stability and continuity than national governments. Their investments are made for the next generation rather than for the next election.

Agencies concerned with the effectiveness, efficiency and sustainability of conservation initiatives can thus profit from people's participation. Participation directly benefits local people as well. When people take part in assessing environmental problems, resources and opportunities, they acquire information and enhance their awareness of the factors that play a role in their lives. When people act and contribute, they often acquire new skills and have the opportunity of organizing themselves, with a variety of returns for local equity, self-reliance and building of community or group identity. However, the very con-

140

cept of participation may be alien to some cultures and groups. For instance, it may be that the self-assertion required to express one's views and interests which differ from those of others is considered "unseemly" and clashes with accepted behavioral norms. The participation of certain disadvantaged groups may clash with local customs, such as the participation of women, the landless, ethnic minorities etc. Also, national governments may not support local participation or empowerment, especially if they regard it as a threat to their own authority.

Moreover, participatory processes require certain investments of commitment, time and resources, and results may take a long time to appear. These resources may not be available or the relevant activities may not have been envisaged in the original plan of the conservation initiative. Also, resource investments are required to reach a good level of communication between the local people and the national or expatriate staff in the conservation initiative. There is no "recipe" for participation, nor any all-purpose description of what it should entail. Nevertheless, every effort should be made to overcome ambiguity, and to be explicit about why, where, when and how people are expected to participate in the conservation initiative. When this is done, it is usually found that certain conditions and forms of support are essential - that participation needs to be allowed, facilitated and promoted. Participation in an environmental initiative will be effective with local people

- assessing their needs and resources, and recognizing the opportunities offered by the conservation initiative,
- taking part in collecting and analyzing environmental and socioeconomic information,
- being consulted on key issues about the initiative (in particular, objectives, design, and key management decisions),
- contributing to planning and decision-making about the initiative at various levels (e.g. local, district, regional, national) which may entail specific negotiation sessions,
- initiating action (i.e. local groups identifying new project needs, and taking action to deal with them, which is different to deciding on tasks identified by project management),
- providing labor and resources to implement the conservation initiative,
- taking part in ongoing decision-making during the implementation of activities,
- assuming specific functions and responsibilities for the conservation initiative, including becoming members of its official management body,

- acquiring benefits from the conservation initiative (this is a poorly recognized form of participation),
- developing effective partnerships with other stakeholders and agreeing on a specific sharing of benefits and costs about the conservation initiative,
- taking part in monitoring and evaluating the initiative.

Tool Box

MOVE

Moderation and Visualization for Participatory Group Events (MOVE) is a tool to elicit ideas and reach group consensus on one or more key issues or courses of action. The exercise needs a skilled moderator who stimulates communication by posing precisely formulated questions on which the group works. Each individual is given time to think and to note down his or her main ideas on cards. The cards are then presented, discussed and grouped to represent the collective reflection of the participants.

Purpose

MOVE is especially useful for planning and priority-setting. Together with other scoring and ranking techniques, the method may also be used when individual opinions must be consolidated into a group decision.

Steps in using the tool

Present the participants with a clear question upon which to reflect. Have the question written on a flip chart or softboard. Give each participant a set of cards and felt pens. Ask them to write down the answers/issues/actions they think are relevant to answer the question. These should be written key words. The participants can use as many cards as they like. Ask each person to come to the front of the group, and read out and explain what he or she has recorded on the cards. As people finish, ask them to pin or tape their cards on the board. The first person spreads cards out. Subsequent people are asked to place their cards close to the ones most similar to theirs or start a new "cluster". When everybody has presented their ideas and placed them on the board, there will be various clusters of items. Ask the participants to consider whether they need to rearrange the cards among the clusters. If they do, they should discuss the moves and agree as a group with the help of the facilitator. The participants may also decide to remove some cards or clusters. Work out together with the participants a title or heading for each cluster. If a rank order is needed among the clusters - for example, deciding which aspects should be followed up - a ranking exercise can be employed. Large groups of participants are broken down into smaller working groups, each to discuss in depth one of the various clusters identified. The smaller groups then report their findings to the plenary session, and a general discussion allows the exercise to be concluded.

Strengths and weaknesses
   +MOVE helps participants group their individual opinions as a collective product,
   +everyone is asked and expected to contribute and the technique promotes paying great attention to the ideas of others,
   +the technique is constructive and adds an important visual element to issues and ideas for action,
   +an "external memory" of the key ideas is produced during the technique,
   – literacy is needed among all the participants,
   – a skilled moderator is essential,
   – a balanced participation of stakeholders is essential.

Ranking Exercises

Ranking exercises are group processes in which participants rank a range of pre-identified actions according to a priority that they assign. The technique follows an assessment process in which people have identified a list of problems and opportunities and/or possible actions to be taken in response to those. It is a particularly good follow-up to a brainstorming exercise or SWOT analysis. The ranking exercise should be followed by identifying, with the participants, the processes required to achieve each of the agreed actions, and by allocating responsibilities for the tasks involved.

Purpose

Ranking is a tool for reaching a group consensus on a course of action to be adopted, and for setting priorities. It can be used when individual opinions must be consolidated into a group decision. Ranking can also be used to identify and quantify needs.

Steps in using the tool

- List the items to be prioritized on a board or sheet of paper visible to everyone. Make the items simple. If necessary, use visual images and drawings.
- Make sure that the participants involved in the exercise are representative of the interests at stake.
- Define a simple ranking mechanism. The system used may depend on the number of items. Where there are more than ten items, each participant can be given a specific number of stickers and asked to stick one or more beside each item they consider important. Where there are fewer than ten items, or where participants wish to weight their judgment of each item, a numbering system may be more appropriate. In that case, each participant allocates a number to each item according to their priority.
- Explain the ranking system to the participants and ask them to think about their preferences and then to place their stickers or write their numbers against the items listed.
- After each participant has ranked the items, compile the group result by counting the number of dots or marks beside each item or by adding up the numbers recorded against each.
- Rank the priorities according to the group's total score and discuss the results with the group. Identify and explore disagreements if any exist.

Strengths and weaknesses

- +ranking is a flexible technique which can be used in a variety of situations and settings,
- +decisions about what should be done and the order of priority are made by the group as a whole rather than being imposed on them. Ranking through consensus is helpful in increasing group commitment to a program of action,

144

+ everyone is able to contribute without having to express themselves in a public forum which can be intimidating for some people (e.g. women and vulnerable groups),
+ ranking exercises are generally found amusing and interesting by participants,
+ relatively large numbers of people (up to about 50) can participate in the exercise,
− choices may be affected by highly subjective factors,
− "block voting" by certain groups can bias the result. If this is a potential problem, the participants may need to be carefully selected to ensure that different interests are represented fairly ,
− the course of action eventually decided on may be different to the priorities of the ranking exercise because of factors such as delays in obtaining necessary resources, or because some things can be achieved quickly while others require more time. To avoid misunderstandings, process considerations should be worked through with the group once the items have been ranked.

(see MOVE, forthcoming)

Types of Participation and their Evaluation

There are various ways and degrees for people to participate:

- **direct participation** - people personally express their opinions, discuss, vote, work, offer a material contribution, receive a benefit, represent themselves,
- **semi-direct participation** - people delegate others, e.g. relatives, friends, respected members of their community, representatives of a community-based group, to represent them in all sorts of activities, but maintain a direct, face-to-face relationship with their representatives,
- **indirect participation** - people delegate others, e.g. experts, appointees of large associations, NGOs, parties or government officials, to represent them in all types of activities, but rarely, if ever, interact with their representatives on a person-to-person basis.

As participation is crucial, it is important to know how to measure and evaluate it. Two approaches from Africa and Asia may illustrate this. In a GTZ environmental management project in Burkina Faso three aspects - **1** Influence of the target group on decisions in general, **2** their representation in consultations and **3** in decision-making are evaluated in four fields of decision-making: **A** - Project Identification, **B** - Determining the Terms of References, **C** - Decisions on Local Activities and **D** - Organization of Crucial Tasks. The actual degree of participation is measured on a six-grade scale from "no participation" (0) to "autonomous decision" (5). The resulting "Participation Profile" (see below) makes comparisons on different levels easier - it shows, for example, that project planners have a different perspective from the people concerned regarding participation in determining the terms of reference (Wolff 1992)

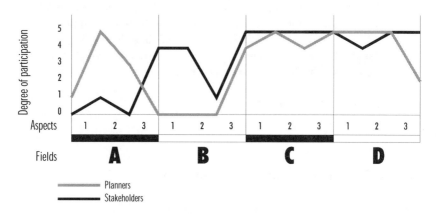

A similar approach was used by ACT - Appropriate Communication in Development in Indonesia (Oepen 1988). The key indicator of participation is who really determines the agenda and the impacts for changes in attitude and practice. According to a set of criteria, the influence of various actors - outsiders, local elite, local motivators, individuals concerned, and groups concerned - is measured in terms of the degree of participation and self-determination in consultations and decision-making. This is applied to various project levels - planning, media and content selection, communication processes and media production, evaluation, follow-up in media and community development. Initiatives, activities and changes in practice are evaluated in terms of the degree of influence by the above actors on a 1o-grade scale from "no involvement" (0) to "autonomous decision" (10). The related chart allows for a quick overview of who has a say in what matters. In the case below, the initiative was originally (level 1-2) an external one but was gradually taken over by local groups (level 3-5).

Levels	Influence by				
	Outsiders	Local Elite	Local Motivators	Individual concerned	Group concerned
1 Planning	7	7	3	–	–
2 Media and content selection	6	6	6	4	8
3 Communication processes and media production	2	4	8	8	8
4 Evaluation	7	3	7	2	7
5 Follow-up in media and community development	–	–	7	7	9

Media Selection and Mix

Based on the previous results of audience and KAP analyses and the preliminary considerations regarding the participation of strategic groups, an appropriate multimedia mix should be developed. The media selected should be appropriate to the audiences' information-seeking habits, preferred information sources, media access, media consumption patterns, communication networks, and group communication behavior.

An important aspect in employing a multimedia approach is the proper selection of available channels in order to avoid redundant or superfluous media usage and to optimize the level of multimedia support required. Thus, a mul-

147

timedia approach does not mean that all available communication channels should be utilized. General guidelines for selection of the multimedia mix should be based on the specific campaign objectives and strategy, KAP levels, etc. For instance, the degree of emphasis in utilizing mass, interpersonal or group communication channels depends on target beneficiaries' KAP levels and campaign strategy priorities. In developing an appropriate multimedia mix, results of audience analysis should be considered, especially on information-seeking habits, preferred information sources, media access or ownership, media consumption or usage patterns, communication network interactions and group communication behavior.

The rationale is that a coherent, coordinated and reinforcing system of communication should be able to address specific but varied information, attitude and behavior problems and needs of intended beneficiaries. Experience and research show that using a combination of mass, group and interpersonal communication in a multimedia mix is most effective because

- no medium is effective for all purposes or all target beneficiaries,
- a communication strategy usually has various information, educational and communication objectives,
- different media and communication channels complement and reinforce each other,
- strategic planning means to select which medium or combination of media should be used for what purpose and by whom, in order to exchange which specific messages with whom.

A General Approach to a Multimedia Mix

There are of course many different media available which have widely diverging characteristics and qualities, advantages and disadvantages (see table below, also SPAN 1993). They also vary greatly in the requirements that need to be met in order to use them, and in the purposes for which they can be used. In general, select and use a medium

- for a single or specific rather than for different goals,
- that has a unique characteristic or particular advantage which is useful to accomplish a specific purpose,
- which the target audience is already familiar with and has access to,
- which can easily accommodate "localized" messages,
- that can be locally developed, produced and operationally supported,

- that complements and reinforces others used in the same strategy while having distinct functional strengths and emphases.

Also, it is useful to distinguish between media and material. Posters and films are materials, i.e. "the carriers" of your message while walls and TV are the media, "the vehicle" that brings the material with the message to the intended beneficiaries.

Not only mass media are useful in this context. GTZ, for example, bridges intercultural barriers between farmers and consultants in Bolivia and Nepal by means of comics and photo stories with a high degree of identification and mobilization potential. Projects of the forestry sector use street theater for awareness-raising on environmental issues in Honduras and rural radio in local languages in various countries. In Malawi, GTZ realized that modern mass media are not always appropriate to matters of environmental health and, therefore, opted for traditional media (Hollenbach 1993).

Media	Advantages	Disadvantages
Television	prestigious + persuasive potential wide range	monopolized by powerful political + commercial interest groups one-way rarely available in rural areas rural program production rare expensive production + reception difficult + expensive to decentralize
Radio	wide coverage and availability in rural areas cheap production + reception simple and widespread rural program production easy + cheap to localize compatible with oral traditions + local culture	tends to be one-way tends to be 'professionalized'
Video	minimal running costs freedom from processinghandling ease immediate results interaction options	high initial investment dependance on maintainance training requirements technical quality requirements equipment fragility dependance on power + monitor compatibility problems hard to use two-way
Printed Materials books, brochures, magazines, newspapers, ma uals, handouts	mostly cheap, simple and easy to produce can be taken home or copied as permanent reminder supports AV-media	illiteracy of rural populations often absorbed bythe ones who al-ready know (teachers, extensionists) hard to use two-way

Media	Advantages	Disadvantages
Audio-Visual Materials such as posters, flipcharts, foto-stories, wall-papers	mostly cheap, simple and easy to produce easy to transport + flexible to use in different contexts good for training + extension can be used two-way	visual illiteracy of rural populations intercultural misinterpretation of pictoral information pre-testing expensive artists needed
Slides	mostly cheap, simple and easy to produce easy to transport + flexible to use in different contexts good color + visual quality good for training + ex tension	intercultural misinterpretation of pictoral information fotographers + lab needed not to be used in daylight dependance on power + projector hard to use two-way
Audio Cassettes	easy + cheap to produce cassette players + batteries widely available good for information exchange + feedback to + from farmers can be repeatedly sed compatible with rural radio can be used two-way for 'narrow-casting'	lacks visual dimension circulation of cassettes may cause ogistic or social problems needs backstopping system
Traditional Media such as theatre, drama, songs, puppetry, story telling, games, etc.	no capital investment readily available in a variety of contexts, places + times cannot 'break down' technically appropriate to local culture + language highly credible and persuasive 'infotainment' two-way by nature	may lack 'modern' appeal requires some pedagogical and artistic skills needs group organization + coordination usually limited to locality or region
Flanell Pictures such as Flanoflex in Indonesia or CFSME or GRAPP in Africa	cheap, simple and easy to produce easy to transport + flexible to use in different contexts good color + visual quality appropriate to local culture + language good for training + extension can be used two-way	may lack 'modern' appeal requires some pedagogical and artistic skills needs extension system

Cross-cultural Communication and Local Media

In many parts of the world, traditional theater forms and puppets do more than entertain: they convey messages which may at times challenge the *status quo*. Puppets are often able to express ideas and challenge accepted norms of their society. They are a buffer which allows the young to communicate to the elderly issues which are too delicate to address directly. The young can challenge the elderly through the puppets. It may be the only medium that is socially acceptable to deliver such a message. The example is important because it demonstrates the close link between local knowledge, traditional media and

150

social patterns to accommodate change, all of which maintain the delicate balance taking place as communities strive to face new options. Economic and environmental degradation place great stress on affected communities, and local expression often becomes the buffer and the means to present evolving survival strategies.

Tool Box

Street or Village Theater

Street or village theater uses local storytellers, theater groups, clowns, dancers and puppets to inform people about an issue by telling a story. Instead of using theater "for" the people, it can also be used "by" the people themselves, e.g. teenagers supported by school teachers or waste pickers assisted by an NGO (see e.g. the case studies by Trudel and Oepen in PART 5). The presentations use imagery, music and humor to raise people's awareness of an issue that is affecting them. Local people can be encouraged to join in and play a part in the presentations. The presentations can be filmed for TV or recorded for radio and, thus, be made available to a wider audience.

Purpose
To raise awareness of issues by presenting information and possible solutions in an "infotaining" way, closely associated with the local culture.

Steps in using the tool
- If possible, cooperate with local groups that are already working on a local environmental problem. Link them to theater specialists.
- Write up the story line, script and message in close cooperation with community members. Review the message on the basis of their comments.
- Meet with local artists to discuss how the message could be incorporated in the story and told through a play, dance or some other local form of entertainment.
- Support the production of a show, and have it tested with a small local audience for interest and effectiveness.

- Plan to stage the show at a gathering in the area, such as a festival or market day. Presentations can also be taken to schools, and events can be "created".

Strengths and weaknesses
+ an entertaining and non–threatening way of putting across a message,
+ based on local customs, traditions and culture and therefore readily understood and accepted,
+ does not require large capital investment,
+ does not depend on technology that can break down,
+ can be highly credible and persuasive where traditional media have a strong tradition,
− requires drama skills in crafting environmental messages into the theater fabric,
− may be difficult to organize and requires a close working relationship between environmental activists and media artists.

Bridging Local and Outside Knowledge

Bridging local and outside knowledge is the challenge of cross-cultural communication. Communication happens when information from one person's knowledge base is packaged and exchanged with another person's different knowledge base in a form that they both understand, and from which they both can derive mutually enriching meaning. The "packaging" and the "transfer" of the information can take place in many forms, ranging from traditional media via simple graphics and illustrations as used in PRA to electronic media such as radio, video or TV. Traditional media works best among people of the same culture, as the information tends to be coded according to accepted symbols and perspectives. An example is humor, which, even as an effective tool to enhance communication, loses its value easily across culture: translating jokes between languages is rarely possible. A good facilitator learns to decode traditional media, much in the way a social anthropologist learns to understand a culture. In selected instances, the facilitator seeks to harness the creativity of local artists to convey messages from outside. This approach, however, requires extensive investigation into local perspectives to ensure that the message is understandable and meaningful locally.

Visualizing Information

Another dimension of learning about local knowledge is the visualization of local knowledge using simple graphics and maps. Participatory rapid appraisal (PRA) refers to a number of tools by which a facilitator may assist rural communities to visualize their knowledge and share their understanding of their environment through information presented as drawings, models and maps (see Step 1 in PART 4). Using media which locals and outsiders can employ with the same ease and under the same conditions is an important contribution to dialogue and understanding. A simple diagram establishes a common language. Visualizing communication networks is another useful entry point for the environmental planner. This PRA tool allows a community and a facilitator to map out the networks of information exchange which all stakeholders exploit, and it helps identify sources of information and major patterns of communication. Audience segmentation based on KAP (step 2) and media selection (step 6) as well as message design (step 7) are closely interrelated in the example below from the "Pest Management" project in Thailand mentioned before (see FAO 1994, Adhikarya, 1987).

Tool Box

Audiovisual Presentations

Audiovisual presentations can be made with slides or filmstrips accompanied by live comment or an audiotape, or with videocassettes. Filmstrips are made by printing slides onto a film. A filmstrip projector is used and the accompanying dialogue can be recorded onto a tape or read from a prepared script by the presenter. Slides are more cost-effective than filmstrips if the program needs to be changed to suit different audiences, or if the slides will only be used occasionally. Where the same set of slides needs to be shown many times, or several copies are to be distributed, or the presentation needs to be carried around to several areas, filmstrips are more efficient. The filmstrip projector can be powered by a lightweight solar rechargeable battery which enables the images to be shown even in areas where electricity is not available. If making a filmstrip is not possible, slides can be recorded on a videocassette together with the sound track, which at least makes them easier to transport.

Purpose

Audiovisual presentations can be used to promote ideas, teach techniques or stimulate discussions among the people in an area affected by an environmental initiative. They can also be used to inform decision-makers and regulators about how the local people view the environmental issues confronting them. Audiovisuals can be suited to both small and large audiences by simply adapting the size of the projected image.

Steps in using the tool

- Prepare a script, for example a "story" about a local environmental problem, how and why it is happening and how it could be fixed.
- Plan each picture to match the text. Show images that are familiar to the audience and, if appropriate, show images of possible solutions.
- Feature people with whom the audience can identify. If using sound recording, let these people talk about problems and solutions in their own way.
- Before showing the presentation, introduce the topic to the audience, possibly bring up a question that the audiovisual might help the audience to answer. Encourage the audience to make comments throughout the presentation.
- If the audiovisual is not specifically about the local area, make comments and ask questions throughout the presentation to make sure that the local people understand how the situation shown relates to their situation.
- After the projection, summarize what has been seen or ask the audience to do this. Encourage the audience to discuss what they saw. Invite them to approach the screen to identify objects, ask questions, give advice, etc.
- Check if people need more time to understand the material included in the presentation. Some groups may need to see the presentation more than once before they feel confident about discussing it openly.
- Audiovisuals can be used for interviews with local people (farmers, community leaders, etc.) where they talk about environmental problems, needs and concerns. Photos or slides can be used to show what they are describing.

Strengths and weaknesses

+ the combination of image and dialogue allows people to absorb a relatively large amount of information,
+ the novelty of audiovisuals encourages people to attend meetings,
+ people can immediately relate to the issues described by local images,
+ the slides can cover an extended period so that people can, in one short session, see what happens over time,
+ problems can be contrasted with visible solutions from other areas,
+ filmstrips can be easily copied, distributed and stored,
+ the sound tapes can be recorded in different languages,
– slide and filmstrip production require laboratory processing that may not be available in some countries,
– preparation of the presentation takes time and needs some prior experience,
– a comparatively large capital investment may be required,
– equipment requires maintenance and safe storage.

Picture Stories

Picture stories can be presented in the form of comics, flannel boards or flip chart drawings or some variation of these. They are illustrations or photos of problems and solutions which can be put in sequence to tell a story, and can be altered and added to in response to community feedback. Flip charts are basically large sheets of cloth or paper with drawings and simple diagrams illustrating particular points. They enable ideas to be presented in a simple, colorful format that creates interest and is easily understood. Flannel boards are picture "paste-ups" which can be attached to various surfaces in many combinations.

Purpose

Picture stories are used as a support for presentations and discussions. They can also be used to stimulate discussion and community input. People can be asked to add to the drawings on the flip charts or to change the layout and content of the flannel boards to illustrate their own points of view or concerns. The simple, colorful pictures can be very effective in helping participants remember the key messages of a presentation.

Steps in using the tool

- Work out the key messages that the conservation initiative wishes to communicate.
- Work out how to show each of those messages in picture or simple diagram form.
- If using flip charts, place the sheets in the order you want for the presentation and fix them to something, such as a large piece of wood, that will enable the sheets to be hung up and turned over during the discussion.
- If using flannel boards, make sure extra materials (e.g. figures of people and animals) are available to allow for the preparation of paste-ups of issues and solutions raised by the group.
- Field staff should be trained in how to use the flip charts or flannel boards to stimulate discussion and to help the participants reach decisions.
- Field-test the illustrations with a few of the local people to make sure the intended messages are understandable. Use questions like "What do you see in this picture?, What does the scene say to you?, How might we change this picture to show the message better?"
- During presentations, encourage the audience to join in with questions, answers and points of view. If using flannel boards, participants should be involved in putting the paste-ups on the board and moving them around in response to feedback. With flip charts it is probably better to use sheets of paper rather than cloth if the audience is going to be encouraged to add to the sheets.

Strengths and weaknesses

- +messages can be shown completely in picture form so literacy is not required,
- +equipment is easily transportable,
- +sophisticated technology is not required, and the charts cannot break down, they are ideal for rural situations,
- +equipment is cheap to produce,
- +if smaller versions are made of the charts, these can be photocopied and distributed to participants to take home as a reminder of the discussion and to spread the message,
- +cloth sheets are durable for field conditions, resistant to tearing, heat, dust and rain,

- particular skills are needed to illustrate issues in this way,
- the issues may be too complex to be fully explained in this form,
- format can limit spontaneity and two-way communication unless done in a way which allows the group to interact freely,
- explaining the images with some local people is essential to avoid the risk of miscommunication: After a malaria education program, people felt they did not have to worry about the local mosquitoes as "We never had mosquitoes around here as large as those on the chart we saw today!" (see Fuglesang, 1982).

Radio Programs

Radio programs can be a useful tool to inform people in a large area. They can be produced at the local, regional or national level. They are most effective when they are made with audience participation in the local language and take cultural traditions into account. Radio production teams should be multidisciplinary and mobile so that they can converse with a range of people and record a variety of material in various locations. Specific programs can vary from formal documentaries to discussion forums with a range of local actors, from plays and storytelling to talk shows where people can phone in and express their views on the air.

Purpose

Radio programs can be used to spread information, to stimulate discussion and debate among the people concerned about the conservation initiative, or to provide a forum where rural communities can communicate their views to others in the region. They can also help to educate and inform decision-makers and regulators, both within and outside of the area, about how the local people view the environmental issues confronting them. Issues raised can be addressed immediately, or subsequent broadcasts can have technical staff and decision-makers answer questions raised by local people.

Steps in using the tool
- Identify a radio station willing to host a program on the conservation initiative. Establish an agreement with the station, possibly on a

regular basis and at a popular listening time.
- Have some staff trained in the techniques of preparing a radio program, including interviewing.
- If a documentary approach is to be used, prepare a story line but, as far as possible, involve local people in designing the program.
- For interviewing, select local people who are able to express their ideas and experiences clearly and who can present a range of experiences and perspectives.
- Edit the tapes so that they present a coherent picture of the issues confronting the community and the conservation initiative.

Strengths and weaknesses

+can inform many people over a wide area within a short time,

+can strengthen the sense of community and of shared experience,

+if aired on a regular basis, radio programs can be invaluable as a forum for discussion on the conservation initiative,

+tapes can be copied and distributed to organizations and schools to use as a focus for group discussion,

– relies on people having access to radios or to telephones (for talk shows),

– cost and time involved in preparing documentary programs are substantial,

– use of recording and editing equipment requires technical knowledge,

– can only be used for raising awareness, not as a substitute for face-to-face discussions with the affected community and other stakeholders.

(adapted from: IUCN 1997)

STEP 7

Message Design

The effectiveness of a communication strategy largely depends on the ability of its messages to catch the attention and understanding of the target audience. Therefore, messages must be designed to suit the specific characteristics, educational and intellectual horizon and the aspirations of each group of intended beneficiaries. Also, they should fit the media selected. This is why they should not be formulated early on in the strategy development. Otherwise, *one* project's message may contradict *another one's*, e.g. "establish a fish pond" by the nutrition campaign may be counteracted by "get rid of non-running water" by a health campaign.

Message Effectiveness

For the message to be successful, the information should be accessible, accurate, verifiable, complete, timely and relevant. Message effectiveness (M.e.) is a function of the reward (R) the message offers and the efforts (E) required to interpret and understand it, hence M.e. = R : E. This concerns both the textual and the visual or audio information in a message. For example, if a poster promises collecting recoverable materials such as paper and plastic from waste as an economically attractive reward (R) while the graphics do not support this vision easily (E) , the message effectiveness (M.e.) will be low. Therefore, an option that is within the control of EnvCom planners is to decrease the level of effort required by the intended beneficiaries in interpreting, perceiving and understanding the messages designed.

Positioning a Message
As especially urban populations are burdened with an "information overload", messages need to be strategically "positioned" so that they "stand out" from the others. They may otherwise not be noticed even though they are relevant and useful to the target audience. The positioning of a message should ensure validity and relevance and outline the general strategy approach, e.g. informational, motivational or action-oriented (see If-Then Chart in Step 4).

Furthermore, it should identify a message focus or theme according to the strategy's issue or objective, and should make the theme attractive and persuasive by "packaging" the message utilizing psychological or social appeals such as

- fear arousal
- incentive/rewards
- testimonial
- morale-boosting
- common man
- guilt feeling

- authoritative
- emotional
- civic duty
- community or peer group pressure
- role model

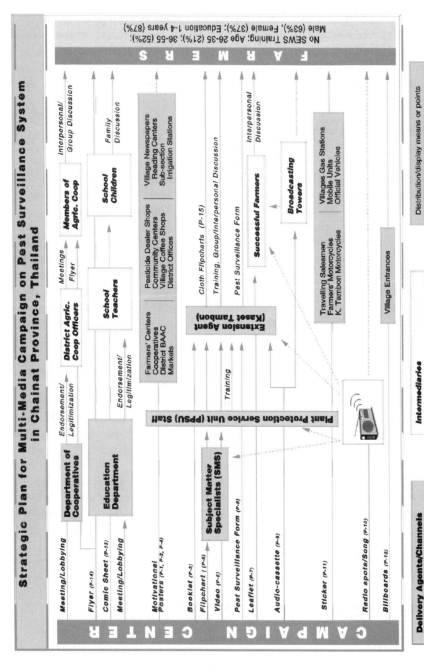

Strategic Plan for Multi-Media Campaign on Pest Surveillance System in Chainat Province, Thailand

Audience segmentation, media selection and message design

160

The theme should be given a special treatment in line with the strategy's objectives, e.g.

- serious/formal
- humorous/direct
- one-sided
  (i.e. pros or cons only)
- conclusion-drawing

- aggressive/confrontational
- popular/informal/indirect
- two-sided (i.e. pros and
  cons) fact-giving
- repetitive

The positioning should take advantage of the specific strengths and potential of the various media selected, e.g. visual media for fear arousal and emotions or print media for fact-giving and conclusion-drawing. It is essential for cost- and time-saving reasons to pretest messages carefully for each type of media and each group of target beneficiaries, especially as far as visual information and (semi-) illiterate beneficiaries are concerned.

Factors in Message Design

There are two main types of factors to consider in message design, i.e. audience and format/content factors. Some of the crucial questions addressing those factors are the following:

Audience Factors   What other ideas, products or practices do the intended beneficiaries have that are different from the ones to be promoted? Which specific needs for each of the segmented audiences does the practice or idea promoted fulfill? What specific questions and doubts are in the audience's mind about the promoted practice or idea? Which aspects of the practice or idea should be emphasized in the message? How should the practice or idea be positioned in the audience's mind?

Format Content Factors   Which words do the various audience segments use to talk about the practice or idea to be promoted? Who should be the source of the message? How will you deal with the numerous factors identified in communication research, e.g. order of presentation, one- versus two-sided appeals etc.? What is the ideal length and pace of the message? What is the appropriate level of complexity, i.e. graphical and textual comprehension?

Audience Participation-Based Message Design

As media production and use costs money, time and skills, the efforts are a worthwhile investment only if communication is successful. Without systematic audience involvement from start to finish this is hardly possible (see e.g.

Rice/Atkin 1989). There is a very simple argument why this fact should be recognized: Whether the media are state-controlled or privately owned, the audience decides whether and which poster, radio program or newsletter will communicate. Therefore, the needs and preferences of the audience should determine how EnvCom planners and producers design media messages - or, as a German saying goes "The worm has to be tasty to the fish, not the fisherman". The recommendations below, while somewhat overlapping with earlier and later steps of the EnvCom strategy, outline a systematic community-centered approach to audience participation-based message design (Mody 1991):

- Learn everything about the topic at stake.
- Observe the life-styles and values of different segments of the audience for decisions on how to communicate.
- Engage in a dialogue with the audience on what each segment already knows, feels and does in relation with the promoted idea or practice.
- Write down objectives: What audience impact should be used to measure whether communication has been achieved?
- Determine which channels and what frequency of exposure are required to reach the objective set.
- Design a creative and persuasive strategy to package the messages to be shared.
- Write specifications for every message, describing its goal, content, and recommended format and treatment.
- Pretest the strategy on a sample of the audience to find out whether the chosen approach is working.
- Modify the message design according to pretest findings and then proceed with final production.
- Monitor physical exposure, attention, comprehension and implementation levels after message distribution or media use begins.
- Evaluate whether the message is achieving its knowledge, attitude and practice (KAP) goals.

The checklist below from a "Pest Management" project in Thailand clearly indicates how message design and media selection are closely related to earlier stages of the communication strategy, e.g. problem identification or KAP analysis. Making maximum use of cost-effective ways of communication may start from a detailed assessment of the most appropriate media in relation to preferences of carefully segmented audiences. Whether or not all media are employed later on, largely depends on the financial and human resources available to the respective project (see e.g. FAO 1994, Mody 1991).

MEDIA TYPE	MAIN MESSAGE	TO SOLVE PROBLEM	FOR WHOM
Motivational poster A	1. Spiders kill planthoppers 2. Excessive use of pesticides will destroy spiders	1, 2, 6, 7	Farmers
Motivational poster B	1. Check your field planthoppers: spray only if you find 40 hoppers in 4 plants or hills, and no spiders	4, 7, 8	Farmers
Booklet	1. Identifying pest 2. Simplified technology on pest surveillance 3. Use of resistant varieties 4. Safe use of pesticides	1 - 9	Extension Agents
Motiv. poster C	1. Safe use of pesticides	1 - 9	Farmers
Flipchart	1. Use of resistant varieties, identification of pest and natural enemies, steps in pest surveillance, safe-use of pesticides	1 - 9	Extension Agents
Video	1. Identification of natural enemies 2. Surveillance and Early Warning System 3. Safe use of pesticides	1 - 9	Extension Agents
Leaflet	1. Steps in pest surveillance 2. Use of resistant varieties	1 - 9	Farmers
Pest Surveillance (PS) form	1. Use of simplified PS form 2. Importance of checking fields properly 3. Use correct Economic Threshold Level (ETL) 4. Use right chemicals	1, 5, 7, 8	Extension Agents, Farmers
Audio-cass. tape	1. Motivational radio spots and songs	1 - 9	Ex.Agents, Farm.
Radio spots and songs	1. What is pest surveillance? 2. Proper disposal of pesticide containers 3. Spray only at correct ETL 4. What natural enemies can do 5. Use of resistant varieties 6. Going into the field is easy, will not destroy plants	1 - 9	Farmers, Extension Agents
Sticker	1. Motivation to go into the fields to check	4, 7	Farmers
Billboard	1. Motivation to go into the fields to check	4, 7	Farmers
Comic sheet	1. Result of farmer checking field from dike only	7	School children
Flyer	1. Identifying natural enemies 2. Use of resistant varieties	2, 3, 5, 6	Farmers
Cloth flipchart	1. Steps in a simplified technology in pest surveillance	1 - 9	Farmers

Media - Message - Audience Checklist

Media Pretesting and Production

Management Planning

Comprehensive and detailed management planning is an integral and vital part of the EnvCom strategy. It will not only spell out the implementation procedures and requirements but will also be used to develop a management information system, including monitoring and supervision procedures. The media or material selected should not be mass-produced too early in the elaboration of the EnvCom strategy. The implementation of a multimedia communication strategy has a higher chance of being successful if:

- the media materials are produced as planned and on time,
- the various media are mobilized and coordinated as suggested,
- all actors involved in this process have been trained accordingly, if necessary,
- the impact and effects of the strategy's implementation are assessed by means of a built-in formative (continuous) and summative (ex-post) evaluation.

In general, the following procedures should be taken into consideration. A clear briefing of all media designers and producers on communication materials regarding the content, design, persuasion and memorability should be ensured. A precise production plan for each material should be drawn up in order to ensure that every relevant staff member knows what to do at exactly the appropriate time. Therefore, all staff should be informed as to their involvement and timing as precisely as possible. Special briefing and training for all personnel who are involved in EnvCom activities must be undertaken to ensure that they understand their specific tasks and responsibilities and have the necessary skills and support materials to perform these effectively. External communication experts must be selected for specialized tasks and briefed as to their work and schedule as well.

Pretesting

As media production is a very expensive business and the effectiveness of the messages is absolutely essential to the success of the EnvCom strategy , thorough pretesting should be undertaken before larger quantities of material are produced. The pretesting should take place on location and with a representative sample of those social groups who will later be expected to change their practices in accordance with environmental principles or ideas. Also, it should be defined precisely what should be tested, e.g. relevance, textual and visual understanding, motivation and action potential, acceptance and credibility etc. For practical, relevance and cost reasons, the actual media should be produced as close as possible to where they will be employed. A production timetable should be determined as precisely as possible.

STEP 9

Media Performance and Field Implementation

Management Information System

This is the point in the strategy process where *management planning* takes over from strategy development as the main task of a communication specialist. One of the worst problems in communication strategy implementation is the untimely delivery or even unavailability of inputs or services required for the adoption of the recommended practice changes or actions by the target beneficiaries who have been motivated and persuaded beforehand. This may lead to frustration among members of this group and ultimately undermine the credibility of the strategy. Therefore "Be ready when the people are" should be a primary concern for any EnvCom planner.

To transform EnvCom strategies into EnvCom activities, management objectives must be identified clearly to include at least the following elements: what the action is, who is to carry out the action, how the action is to be carried out, the volume of resources needed and how to obtain such resources, and when the action is to be accomplished. In addition, management objectives should set a standard for measuring the progress and impact of implementation.

Recommended Action

The implementation of a multimedia communication strategy requires a good management information system that provides the organizers with rapid feedback on important strategy activities and thus helps to readjust or change the strategy if necessary. This information system should also take care of the proper coordination of various activities, which often need to be carried out simultaneously. Proper implementation of activities within the estimated time period is also essential. A delay in one of the interrelated multimedia activities will often trigger chain-reaction effects. Realistic time estimates should therefore be accorded due attention.

For the actual field implementation, a timetable for each type of media and each social group should be determined. Also, the most appropriate events, occasions, times and places should be carefully considered with a view to co-ordinating with mass media inputs, if ever possible, and reinforcing the strategy by means of side effects, incentives and non-economic benefits, e.g. games, lotteries, contests, street festivals etc. It is usually advisable to "cross-fertilize" various media and communication channels, e.g. the emotional appeal of radio with the factual impact of print media. The same can be said in relation to planning for multiplier effects among the various media used, e.g. a radio show about a people's theater performance. Events that "stage" media inputs, e.g. festivals, contests, VIP visits etc. can and should be created, if necessary. If possible, "piggy-backing", i.e. getting a free ride on existing communication channels, extension services or other institutional outlets should be encouraged and facilitated. Usually, this calls for building strategic alliances with other institutions and social groups during earlier stages of the EnvCom strategy (see Step 2).

STEP 10

Process Documentation and Monitoring and Evaluation

Evaluation should be made a continuous effort of communication planners at all stages of the strategy. Its major focus should be on the efficiency of program implementation, the effectiveness and relevance of an activity or overall program, and the impact and effects of an activity or overall program.

Types of Evaluation

There are many types of evaluation at various stages of a project or program:
- Ex-ante (appraisal) as part of planning to estimate what effects should be expected
- Ongoing (monitoring) during implementation to assess whether the p rogram is on course (also called formative evaluation).

166

- Ex-post (impact assessment) soon after implementation to ascertain the effects (also called summative evaluation)
- Terminal (impact assessment) some time after implementation to rate the sustainability of effects.

Whereas formative evaluation findings are often utilized to improve the EnvCom strategy or performance during its implementation, the results of summative evaluation are normally used to determine whether the strategy has accomplished its objectives and whether an improved or expanded intervention should be undertaken as a follow-up program. To ensure that a summative evaluation is conducted properly and that its results are relevant to the EnvCom strategy objectives, related activities should be considered as a built-in component and an integral part of the strategy process. The findings of such a summative evaluation should be used as inputs to formulate new or improved strategy objectives, or to help set up new baselines or benchmarks for future EnvCom interventions of a similar nature. It is the information feedback resulting from the summative evaluation which completes the "loop" of the strategy planning process by feeding in relevant evaluation findings back to Step 1 of this process.

Impact Assessment

The impact assessment of an EnvCom strategy may focus on the following parameters and questions:

1 Problem and Research
- Whose problem is being discussed?
- How relevant is it to the audience?
- Is the topic well understood: causes, dynamics etc.?
- What is the total context of the problem?
- Do research results reflect reality?
- Does the problem generate emotion: interest, anger etc.?

2 Choice of Media
- How appropriate is the media choice given the audiovisual literacy of the audience?
- Is there an information overload or insufficiency?
- Does the media choice help to strengthen the message?

3 Effects
- Is the message oriented towards people, not projects?
- Does the media choice respect the culture and sensitivity of the audience?
- Does the message boost self-confidence and self-help?
- Is the message and its delivery non-patronizing and non-propagandistic?
- Are the messages heard, understood and accepted, and, most importantly, do they motivate and mobilize for behavior change and action?

Process Documentation

Based on a chronological description and analysis of successful and less successful decisions made during planning, implementation and management, certain generalizations can be suggested that are valuable in designing or planning future campaigns which have similar objectives. This type of process documentation of the critical issues and decision-making requirements should be started from the very beginning. Regardless of the outcome of a process documentation and summative evaluation exercise, the lessons learned - both positive and negative – are of great interest and value. Therefore, efforts to disseminate and share the summative evaluation results of an EnvCom strategy among EnvCom planners, managers, and trainers should be encouraged and pursued.

Tool Box

Community-based Environmental Assessment

Community-based environmental assessment provides a community perspectives on the state of the environment, prior to or during a conservation initiative, as part of a monitoring or evaluation exercise. A list of environmental aspects or factors is agreed upon by the community. The state of each factor is determined by allocating a certain value (e.g. excellent, good, poor etc.) or number to it. It is not the actual value or number that is important but the way it changes over time as recorded by ongoing observations.

Purpose
Community-based environmental assessment provides a framework within which insiders can make observations and judgments about the state of certain environmental factors.

Steps in using the process
- In a meeting with the community members concerned, discuss the purpose of the assessment and how it can be carried out.
- Decide what is to be measured (e.g. well-being of the community, well-being of a particular natural area) and define what indicators will be used (e.g. abundance of specific species in the area, pollution, soil erosion, migration, morbidity and mortality, wealth, literacy, clean water etc.).
- Write up the values to be used and what each represents (e.g. 5 = very good, 1 = very bad).
- Draw up a list of all the items to be evaluated. If the group is small (less than ten) work through the list together to reach a consensus on what value should be attributed to each item at the present time. If the group is larger, divide into smaller groups, with each group having the same list of items to evaluate. Then bring the groups together to negotiate a common list of allocated values. Record and store the results and decide when the exercise will be repeated.
- At the agreed time, repeat the exercise of assigning a value to the items to be assessed. Discuss the reasons for the values attributed and the causes of changes since the previous exercise.
- Identify the actions which need to be taken in response to the analysis and who should take responsibility for each task.

Strengths and weaknesses
+ enhances local knowledge of environmental issues,
+ creates an awareness of the potentially negative and positive environmental impacts of activities,
+ fosters the development of evaluation skills among participants,
– this is quite a complicated tool and a clear explanation is required to make sure it is well-understood before assessment begins,
– some value allocations may be highly subjective, although discussing the reasons for the allocations can help reduce and clarify this,
– it may be difficult to reach consensus on "values" where there are significant disparities between the costs and benefits experienced by different stakeholders in relation to the relevant item.

The evaluation results from the "Rat Control" Campaign show the changes in terms of the KAP levels of rice farmers in Penang vis-à-vis rat control campaign recommendations and messages. As a result of the campaign, the number of farmers who reported that all the rice plant damages were due to rats fell from 47 percent before the campaign to 28 percent after

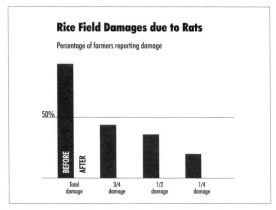

Evaluation results from the 'Rat Control' Campaign

the campaign. The rice field damages due to rats in 1984 (before the campaign) covered about 700 ha compared to only 223 ha in 1988 (FAO 1987-89). The economic cost-benefit ratio of the "Rat Control" Campaign as a whole is presented overleaf.

Cost and Benefit Analysis of Malaysia's Rat Control Campaign (In US $)

Acreage loss [1]:	1984 =	700 ha	or	1,729 acres
	1988 =	223 ha	or	551 acres
Production estimates:	1 acre =	1,600 kg [2] =		$310
Financial loss:	1984 =	2,766,400 kg =	$ 535,990	
	1988 =	881,600 kg =	$ 170,810	
Difference =		1,884,800 kg [2] =	$ 365,180	
Total savings =			$ 365,180	
Total expenditure for campaign =			$ 140,184	
Cost/benefit ratio		**1 : 2.60**		

For each $ 1 invested, a return of $ 2.60 was gained.

Campaign Target:	14,000 farm families	
Average Economic Benefit per Farm Family:	**$ 365,180 : 14,000 = $ 26**	

PART 5 - Case Studies

Environmental Activism Through an Entertainment-Education Radio Soap Opera in India

Arvind Singhal, Saumya Pant and Everett M. Rogers

In the past decade, a "new" communication strategy called entertainment-education has shown promise in addressing such social problems as unchecked population growth, gender inequality, environmental degradation and illiteracy (Singhal/Rogers 1999, Piotrow et al 1997). Entertainment-education is the process of purposely designing and implementing a media message to both entertain and educate, in order to increase audience members' knowledge about an educational issue, create favorable attitudes, and change overt behavior (Singhal & Rogers 1999). Entertainment-education seeks to capitalize on the popular appeal of entertainment media to show individuals how they can live safer, healthier and happier lives, and to show communities how they can address such problems as large family size, environmental pollution, HIV/ AIDS, and the like.

The idea of combining entertainment with education is not new: It goes as far back in human history as the timeless art of storytelling. For thousands of years, music, drama, dance and various folk media have been used in many countries for recreation, devotion, reformation and instructional purposes. The conscious combining of entertainment appeals with educational messages, however, in radio and television soap operas, comic books, and rock music is a matter of the past 25 years (Singhal and Rogers 1999).

The present case describes the Indian experience with an entertainment-education radio serial, *"Yeh Kahan Aa Gaye Hum"* ("Where Have We Arrived?"), which promoted environmental issues in India, and, among other things, helped spur environmental activism in Village Lutsaan of India.

"Yeh Kahan Aa Gaye Hum"

In 1998, All India Radio (AIR), the Indian national radio network, under the leadership of Mrs Usha Bhasin, a well-known radio producer and director, broadcast a highly popular entertainment-education radio serial entitled *"Yeh Kahan Aa Gaye Hum"* ("Where Have We Arrived?") to explicitly tackle environmental issues. Through an entertaining storyline, the 52-episode radio serial addressed environmental issues such as air, water and noise pollution, deforestation, solid waste disposal, organic farming and other topics.

In designing *"Yeh Kahan Aa Gaye Hum?",* Mrs Usha Bhasin and her team capitalized on lessons learned during the past decade in producing several highly popular entertainment-education radio serials dealing with such issues as adolescence (*"Jeevan Saurabh"* in 1988), marriage incompatibility (*"Jeevan Saurabh II"* in 1989), teenage sexuality (*"Dehleez"* in 1994-1995), and women's status, family size, and community harmony ("Tinka Tinka Sukh" in 1996-1997) (Bhasin and Singhal 1998, Singhal and Rogers 1999). The environmental issues promoted in *"Yeh Kahan Aa Gaye Hum"* were culled with the assistance of the Central Pollution Control Board of India.

India's renowned music lyricist, Javed Akhtar, wrote the catchy theme song *for "Yeh Kahan Aa Gaye Hum",* which was sung by the highly popular singers Kavita Krishnamurthy and Shankar Mahadevan. A great deal of preprogram publicity over several months preceded the first broadcast of *"Yeh Kahan Aa Gaye Hum".* This preprogram publicity was carried out via radio, television, and the national, regional, and vernacular press. In addition, some 60 non-governmental organizations working on environmental issues in the broadcast area were involved in promoting the radio serial in their local area of operation. Representatives of these organizations met in Delhi for a two-day workshop organized by Mrs Usha Bhasin to chalk out key environmental issues, especially the ground realities of how these issues could be acted upon by listeners. The program was sponsored by Breeze and Lifebuoy, two well—known brands of the Hindustan Lever Company, providing an opportunity for AIR to recover its production costs.

The epilogues in the serial were delivered by Shabana Azmi, a popular Indian film actress, a social activist, and a member of India's parliament. An epilogue is a concentrated 30 to 60-second advertisement for the educational message, usually delivered by a credible media celebrity. Following the broadcast of each episode, Azmi summarized the modeled messages about pro and anti-environmental behavior, goading listeners to launch village-clean up drives, plant trees, conserve water, and use reusable cloth bags instead of disposable plastic ones. She also encouraged listeners to write to AIR describing their perceptions of the positive and negative environmental behavior depicted in the program, and specifying what new forms of behavior they had incorporated in their day-to-day lives. A competitive spirit was fostered among listeners by awarding prizes for the quality of the feedback they provided. Outstanding community work by listeners in the realm of environmental conservation was also recognized.

"*Yeh Kahan Aa Gaye Hum*" was broadcast from June to December 1998 via 31 radio stations of All India Radio, covering seven Indian states in the densely populated Hindi-speaking area of northern India: Uttar Pradesh, Bihar, Madhya Pradesh Rajasthan, Haryana, Himachal Pradesh and Delhi. Some 600 million people comprising 100 million households live in these seven Indian states. Our previous research suggests that such entertainment-education programs can attract a regular listenership of 40 to 50 million people (Singhal and Rogers 1999), which, in the case of "*Yeh Kahan Aa Gaye Hum*" may constitute the largest audience for an entertainment-education program on the environment anywhere in the world.

Community Effects of Radio in Lutsaan

In addition to analyzing the storyline, epilogues and several hundred listeners' letters of "*Yeh Kahan Aa Gaye Hum*", we investigated the effects of the radio serial in Village Lutsaan in the Uttar Pradesh State of India, where since 1997, we have been studying the community effects of radio. Our interest in Lutsaan was piqued in December 1996, when a colorful 21-inch by 27-inch poster-letter-manifesto with the signatures and thumbprints of 184 villagers, was mailed to All India Radio in New Delhi, then broadcasting an entertainment-education soap opera "*Tinka Tinka Sukh*" (Happiness Lies in Small Things), the predecessor to "*Yeh Kahan Aa Gaye Hum*". The poster-letter from Lutsaan stated: "Listening to "*Tinka Tinka Sukh*" has benefited all listeners of our village, especially the women. Listeners of our village now actively oppose the practice of dowry — they neither give nor receive dowry."

We visited Village Lutsaan for the first time in the summer of 1997 to gauge the effects of the radio serial "*Tinka Tinka Sukh*". Intrigued by this community's voracious appetite for radio listening, and impressed by its unique ability to pop pop radio's educational messages into concrete community development initiatives, we returned to Lutsaan again in July 1998 and January 1999 for extended field visits. As "*Yeh Kahan Aa Gaye Hum*" was broadcast from June to December 1998, our previous two visits to Lutsaan, especially, provided us with a unique opportunity to investigate the community effects of this environmental radio serial. During these visits, our research team spent over 50 person-days in Lutsaan, conducting over a dozen focus-group discussions, two dozen in-depth interviews, photo documentation, and various forms of participant observation activities.

Village Lutsaan, located 30 kilometers from Aligarh, the nearest city, rises some 90 feet above the surrounding Gangetic Plain. Most of the village's approximately 1,000 homes are located on the sides of a small hillock, topped by an ancient fort. The adobe homes neighbor in a dense manner. The population of Lutsaan is about 6,000. Numerous narrow paths meander among the mud houses. Buffalo are everywhere, and the smell of dung pervades the village. Fertile, flat fields surround Lutsaan, and its farmers travel out to work on them each day.

Compared to most Indian villages, Lutsaan is relatively well-off, with 60 radios, 5 television sets, and 10 tractors. Nearly every household owns a bicycle and 25 households possess a motorcycle. Lutsaan has two village schools that offer eight years of education. The village has a Shyam Club with about 50 active members. It carries out various self-development activities, including village clean-up, fixing broken water pumps, and reducing religious and caste tensions in the village. The village postmaster is chair of the Club. He told us that when an interpersonal conflict occurred recently, members of the Shyam Club met with the disputants until a solution was mediated. In 1996-1997, stimulated by the radio serial, "*Tinka Tinka Sukh*," the Shyam Club devoted its main attention to such gender equality issues as encouraging girls to attend school, and opposing child marriage and dowry. Hence, even before the broadcasts of "Yeh Kahan Aa Gaye Hum" began, the village community of Lutsaan was already hooked to radio listening.

Not surprisingly, we found that both men and women in Lutsaan listened to "*Yeh Kahan Aa Gaye Hum*". Two radio listening clubs - one for men, the other for women - each comprising about 25 members, were highly active during the broadcasts of "*Yeh Kahan Aa Gaye Hum*". To publicize the serial and promote environmental activism, these listeners' clubs had set up some 50 hand-painted posters and wall signs, encouraging other villagers to listen to the radio program. The club members paid about Rs 15 (less than 50 cents) monthly, which was used to publicize the serial and organize pro-environment activities for the village.

Our observations showed that the members of the radio club listen to the serial collectively. Each episode of "*Yeh Kahan Aa Gaye Hum*" was followed by a highly engaging post-broadcast discussion. The secretary of the male listening club took copious notes while the episode was being aired, and maintained a diary account of the key happenings of each episode. After the episode was over, he typically set the discussion agenda among the listener members, who then interpreted the key environmental lessons in the episode just broadcast. The discussion usually ended by identifying feasible ideas for implementation at the village level.

We learned that the listeners' club in Lutsaan launched several social campaigns to save the environment. Inspired by the radio serial, members went on a bicycle tour to educate the public to conserve fuel, thus saving the environment from pollution. They rode to the nearest railway junction where several auto-rickshaws waited, with their ignitions on, for the train to pass. The Lutsaan "activists" explained to the drivers the hazards of air pollution, encouraging them to switch off their ignitions. The drivers initially thought that the club members were crazy. But the Lutsaan team persuaded them to switch off their idling engines.

Members of the Lutsaan listeners' club also approached heavy smokers in the village, informing them about its health hazards, including the problems of air pollution. To boost their credibility in these environmental missions, they often took "Bapu", a village opinion leader, with them and other volunteers of the Shyam Mandal, a self-help group in Village Lutsaan. This collaborative synergy between members of the radio listeners' club and the Shyam Mandal helped spur and sustain environmental action in Lutsaan.

Another suggestion that the Lutsaan residents gleaned from the serial "*Yeh Kahaan Aa Gaye Hum*" was to plant trees on festive occasions. During one of our research visits, the club members planted trees. Some ten saplings were prepared for planting in different areas of the village. They knew from listening to the radio serial that these saplings would grow into trees that would provide fruits and shade.

Why was "*Yeh Kahan Aa Gaye Hum*" so effective in stimulating environmental activism in Lutsaan? Exposure to this radio serial was higher in Lutsaan than elsewhere in North India. Prior conditions in the village helped magnify the impact of this entertainment-education radio program: active radio listening clubs, group listening to the radio episodes, a highly-respected village leader in the postmaster, and the activities of a village self-help group.

Promoting Sustainable Rural Development Practices through Training of Agricultural Extension Workers in Indonesia

Mariam Rikhana and Soedradjat Martaamidjaja

For more than two decades, Indonesia has earned high marks for its economic development. Its economy has grown at an annual rate of almost 7 %. With a total surface area of 1.93 million sq km, 11.5 million ha of which is cultivated with rice (1995), and the use of innovative agricultural technologies and management, Indonesia has been able to produce enough food to feed its population of more than 200 million. Agricultural development in Indonesia has therefore significantly contributed to the economic achievements. The great strides Indonesia has made in achieving steady economic growth and alleviating poverty and the transitional process from an agricultural to an industrial country have created considerable pressure on its natural resources. Economic development has frequently taken place at the expense of serious environmental degradation. With the rapid growth of economic activities and the intensive use of natural resources, Indonesia now faces the challenge of sustaining its development.

Like other economic sectors, agriculture is now under increasing threat of environmental and natural resources degradation, caused by uncontrolled, excessive exploitation of natural resources and the use of environmentally detrimental technologies. Land clearing for agricultural and industrial development leading to soil erosion and a loss of groundwater, continuous use of intensive chemical fertilizers and pesticides, overfishing, and agricultural and industrial waste disposal are among practices that contribute to reducing the quality of the agricultural environment and resources. This critical situation calls for serious attention and efforts to protect the agricultural resources from further depletion, and to conserve the existing resources for future productive and sustainable utilization. To protect the agricultural environment, a decree was issued to ban 57 insecticides, mostly highly toxic organo-phosphates, from use in rice production, and Integrated Pest Management (IPM) was declared as the national pest control strategy. In addition, the decree called for large-scale training of field workers and farmers in implementing the IPM program. Beside the IPM program, various specific environmental-conservation-based farming programs have been introduced to promote sustainable agricultural and rural development.

Environmental Education in Indonesia

Environmental issues have been integrated in Indonesia's formal education system since 1975, in both general and vocational schools. Priority was given to teachers' reorientation training and production of reference books and instructional materials. The first instructional materials on environment were developed in 1985, and designed as a pick-and-choose portfolio to allow teachers to be flexible in selecting and using the topics most relevant to their teaching subjects. During the last ten years, more specific instructional materials have been published to be employed by teachers at different levels. Environmental education has also become an optional course or a field of studies offered by many colleges and universities.

Furthermore, environmental issues have been incorporated in agricultural education since the 1980s and integrated into the existing curriculum of the basic agricultural extension training for extension workers. The training materials provided, however, did not allow the extension workers to comprehend sufficiently the environmental issues in agricultural sectors.

Agricultural Extension Training System

To serve the farmers, the government of Indonesia recruits extension workers who are categorized into two groups: Field Extension Workers (FEWs) who have graduated from an Agricultural High School for Development (AHSD) and Subject Matter Specialists (SMSs) who have graduated from an agricultural university or college. Being at the front line of the agricultural extension system, FEWs are assigned to perform the following tasks: (1) disseminating useful agricultural information, (2) recommending the most profitable technology to farmers, (3) imparting sound knowledge and technical know-how in applying new technologies, (4) helping farmers obtain easy access to development facilities available in their areas, (5) motivating farmers in achieving self-reliance and self-supporting enhancement through group cooperative actions, and (6) conducting field observation to assess technologies being promoted or implemented by farmers in their respective areas.

As backstoppers of FEWs, SMSs have an important role in bridging research and extension. Their main tasks are to conduct local verification trials and study of agricultural technologies to assist FEWs and farmers in solving problems related to technological application. In addition, SMSs are assigned to teach and train FEWs with respect to new agricultural knowledge and technological

179

skills. Approximately 36,000 Field Extension Workers (FEWs) and around 3,000 Subject Matter Specialists (SMSs) are currently employed to serve the farmers. The FEWs are stationed at the Rural Extension Centers (RECs), and they are supported by four district agricultural services in both legal and technical matters. The SMSs are employed at district and provincial agricultural services, including the District Agricultural Information and Extension Centers (DAIECs), to backstop the research-related field activities of the FEWs. While the SMSs are university graduates, mostly of agricultural disciplines, FEWs are mainly agricultural vocational high school certificate or post-secondary diploma holders.

About 250,000 farmer groups are now in existence throughout the country. A farmer group, which is initiated and formed by the farmers, usually consists of 20 to 50 members and is led by a chairman - referred to as a Contact Farmer - who is elected by and from among the group members. A Contact Farmer is usually a successful, progressive and better educated member of the group, who is viewed as a partner of the extension workers and is informally considered to be a voluntary change agent. The agricultural extension training for extension personnel can basically be divided into two categories: pre-service and in-service training.

Pre-service training
Pre-service training for FEWs is carried out in the Agricultural High Schools for Development (AHSDs), administered by the Agency for Agricultural Education and Training (AAET). A substantial amount of time is allotted in the AHDS' curriculum to topics of agricultural extension. Pre-service training for SMSs is given by agricultural universities or colleges, which are beyond the control of AAET. The knowledge on agricultural extension provided at these institutions is generally very restricted.

In-service training
In-service training for extension personnel can be categorized into two types i.e. short-term training and long-term training. Both training programs can be carried out either in-country or overseas. In-country short-term training is directly coordinated by the AAET through 33 Agricultural In-Service Training Centers (AISTCs) currently operating across the country. Some short-term training programs are conducted in cooperation with universities and other related training institutions.

Long-term training is usually undertaken to gain a degree in higher education: a Master's degree or doctoral degree. Overseas training is normally available under the auspices of partner-country-assisted projects.

Four categories of short-term training are available for extension personnel:

- **Basic extension training**. This training is divided into Basic I and Basic II, each of which takes one month. They can be conducted separately or combined as a sequential package. The main topics covered in Basic I include general principles, strategy, methods, techniques, organization and management of agricultural extension, while in Basic II, the topics cover the diagnostic and problem-solving skills necessary to help the farmers solve their farming and business problems. This basic training is given to FEWs and SMSs within the first two years of their service.
- **Skill training**. This training is divided into Technological Skill and Extension Skill Training programs. Each takes one to twelve weeks depending on the subject matter taught. Skill training covers various skills related to the application of technologies needed by the farmers; it is classified into different levels of qualification depending on the technological competency mastered. Extension Skill Training includes various extension-related communication skills needed for effective teaching in the extension delivery processes.
- **Promotional Training.** This training is applied virtually to all civil servants and is obligatory for their promotion to a certain service level. It lasts three to four months, depending on the level of promotion.
- **Reorientation Training.** This training is designed to provide new orientation for extension personnel who are transferred to other working areas entailing a different commodity or farming system. It is also employed for non-extension workers who are being newly appointed as extension workers. The duration of the training varies, depending on requirements.

Although environmental education has been included in agricultural education since the early eighties, the quality and proficiency in environmental knowledge and skills of both agricultural secondary school and university graduates do not satisfy the minimum requirements to manage agricultural practices which lead to sustainable agriculture. Almost all newly employed Field Extension Workers (FEWs) and Subject Matter Specialists (SMSs) have very little, if any, knowledge and skills with respect to nature/environmental management and practices. This has become a heavy burden for the extension training, which has to cope with the pressing need for qualified extension workers in handling environment-related issues of agricultural development.

Lack of practical and easy-to-learn training/extension materials on nature/environmental conservation and its relation to sustainable agricultural development is also among the major constraints facing agricultural extension. The development of such materials requires efforts supported by adequate funds and expertise. The extension materials produced by the Ministry of Agriculture are very restricted in terms of the variety of subject matter, while learning materials from other institutions, including universities, are even more limited. In addition, many extension workers and field-based officials are not motivated enough to work on nature/environmental conservation, owing to ignorance on the one hand and lack of positive aspirations towards the quality of life on the other. These unfavorable attitudes have impeded the success of agricultural extension service efforts in promoting sustainable agriculture.

The Development of the Environmental Education Training Module (EETM)

In response to the problems outlined above, the Ministry of Agriculture devised an environmental education training module (EETM) in 1992 to be used by all Agricultural In-Service Training Centers (AISTCs). It was developed through the Agency for Agricultural Education and Training (AAET) with the support of the Food and Agriculture Organization (FAO). The purpose was to provide training materials for Agricultural Master Trainers who would in turn train field extension workers (FEWs) on environmental problems and issues, particularly those related to sustainable agricultural and rural development. The EETM was designed as a 17-hour training package to be integrated into the existing curriculum of extension training. The module covers basic knowledge on principles and elements of environmental issues and practical environmental management in relation to sustainable agricultural and rural development.

Strategy and process of EETM development

The modular format was selected as a learning package because it has proved to be effective in stimulating active learning. It is also considered effective for developing learning materials based on learners' needs. In addition, a module is very practical in view of its flexibility for revision and updating the content as well as the methods to suit particular needs. Modular instruction is also an efficient adult educational method. The Environmental Education Training

Module (EETM) is designed to be used by Master Trainers at the Agricultural In-Service Training Centers (AISTCs) who, in turn, train Field Extension Workers (FEWs) as the second level of the training target audience.

The EETM was developed through the following steps :

1 **Selection of the module writers:** To guarantee its reliable applicability and quality, the EETM was prepared by knowledgeable and qualified persons having expertise in the technical content, training methodologies and training module development. Five persons with these skills were involved in the development of the EETM. They were recruited from one of the Agricultural In-Service Training Centers (AISTCs), which was later assigned to conduct the Training of Master Trainers (TOT) on the use of the EETM. The module writers also served as the TOT facilitators.

2 **Participatory workshops:** Workshops were designed to determine training needs, training objectives and content, evaluation procedures and training methods. To assess the training needs of agricultural extension workers and to determine the module objectives, a participatory workshop was conducted involving agricultural trainers, staff of the State Ministry of Environment, the University of Indonesia, Bogor Agricultural University, agricultural extension workers and the module writers. Also, to determine the evaluation procedures, module content and training/learning methods, several workshops were organized with the participation of the module writers, agricultural master trainers, as well as agricultural extension workers. Based on the results of the workshops, the module was written and complementary training media and teaching aids were developed.

3 **Content validation workshops:** To validate the technical content of the module, several review meetings were conducted, attended by the module writers, the Heads of Agricultural In-Service Training Centers (AISTCs), agricultural master trainers, teachers of Agricultural High Schools for Development (AHSDs), subject matter specialists, extension workers, and officials from related institutions, including NGOs. During the validation session, one learning activity covered in the module was presented and, by using a simulation method, the participants were asked to respond as if they were extension workers. The suggestions and recommendations elaborated by the participants were used as inputs to revise the EETM.

4 **The module try-out:** The revised module was distributed to selected training centers to be tried out on the intended target group of the module, i.e. agricultural extension workers. The try-out was conducted as part of the existing training curriculum, with some parts of the module which are

related to the trainees' needs being presented. However, in some training centers the EETM was tried out as a separate training course in which the whole content of the module was presented. The evaluation results of the try-out were also used to improve the module.

5 **Legitimization:** To introduce the EETM on a broad basis, a circular from the Director General of the AAET was issued instructing the Heads of Agricultural In-Service Training Centers (AISTCs) to use and integrate EETM into the curriculum of the existing agricultural extension training and other related agricultural training programs.

The development process of the module was discussed among members of the FAO environmental education network in three FAO regional meetings, and has been adopted as a generic module development process. The content and structure of the module was adapted with some modification by some of the network's members.

Institutionalization of EETM

To institutionalize the EETM in Indonesia, the following measures were implemented :

- **Training of Master Trainers (TOT).** One batch of training of Master Trainers was conducted at the national level, facilitated by the module writers as the Principal Trainers and assisted by 4 Master Trainers who had undergone early training. The first TOT was attended by 36 Master Trainers from 32 Agricultural In-Service Training Centers. The method used in the TOT comprised microteaching activities in which the participants served as training managers and facilitators. The participants were divided into small groups consisting of 5 - 6 persons. Each group was requested to review the training module, and to give necessary comments before they drew up the training plan. Each group had the assignment to conduct a training session for 10 - 15 extension workers using the module. The training was carried out at a Rural Extension Center (REC). During the training, the TOT participants recognized the strengths and weaknesses of the module as a learning strategy.
- **Establishing team teaching.** Upon completing the TOT on the use of EETM, the Master Trainers were requested to establish a team to teach environmental education at their respective working units. The task of the team was to implement and modify, if necessary, the existing module to match the needs of their particular target groups. Furthermore, the trainers

184

were asked to develop other environmental training modules as needed by the field extension workers and other related agricultural personnel. The team was also obligated to submit semiannual reports on the implementation progress and results of the EETM at their respective institutions to the Agency for Agricultural Education and Training (AAET).

- **Distribution of the EETM to other institutions.** In order to spread the use of the EETM, in addition to Agricultural In-Service Training Centers (AISTCs), the module was distributed to Agricultural High Schools for Development (AHSDs), Colleges for Agricultural Extension (CAEs), and other related institutions, including NGOs, at the national as well as provincial levels.

The progress of EETM implementation in Indonesia

It was reported that at each Agricultural In-Service Training Center (AISTC) at least 2 batches of field extension training were conducted annually. Each training session was attended by 30 field extension workers. The EETM was integrated into the existing training curriculum, partly or completely, depending on the requirements and time available. The target groups of the EETM were not only field agricultural extension workers, but also forestry field extension workers, transmigration field workers, and district, subdistrict and village administration officers. Up to December 1996, approximately 4,500 field extension workers had been trained on the EETM conducted at Agricultural In-Service Training Centers (AISTCs). In addition, around 500 field facilitators of NGOs and over 1,000 other related field workers have been trained.

The success of the EETM has inspired the AAET to develop other environmental education training modules which focus more on environment-related agriculture. Two specific modules entitled "Sustainable Agriculture" and "Environment and Population Education Training Module " have been developed. The purpose is to provide advanced training materials to be used by Master Trainers, as the prioritized users who have mastered the first EETM, to train Field Extension Workers on the principles, problems and alternative solutions concerning sustainable agriculture.

Subsequently, the Ministry of Agriculture scheduled the second batch of TOT to be conducted by the end of 1997. The prime participants of this second TOT were to be Master Trainers from Agricultural In-Service Training Centers (AISTCs) located in eastern Indonesia. The first batch of TOT was restricted to those from the western part of the country. A third batch of TOT will target the teachers of AHSDs and CAEs at a later date. In keeping with the increasing concern for environmental issues, the curricula of the Agricultural High Schools for Development (AHSDs) and the Colleges of Agricultural Extension (CAEs) have been reviewed to ensure greater orientation towards the environment. The AAET has committed itself to produce more specific EETMs in the future, to be used by various educational institutions at different levels.

Results and Lessons Learned

The production of the FAO-assisted EETM has inspired the Ministry of Agriculture to undertake follow - up activities. Since 1994 the AAET of the Ministry of Agriculture has allocated funds to reproduce the EETM and to train field extension workers. Within 3 years, approximately 6,000 field extension workers and related officials were trained using EETM, and more than 15,000 copies of the EETM were produced. In addition, AAET is planning to develop more specific, advanced training modules on environment-related issues. Training modules on organic farming and on IPM vegetables are currently being developed. The establishment of team teaching (3-5 persons) on environmental training at each Agricultural In-Service Training Center (AISTC) is a follow-up action of alumni of the TOT. The team is led by Master Trainers who have been trained through FAO-supported or AAET training projects, and assisted by other local trainers who have been trained by the TOT alumni. The EETM has become a model for the development of other environment-related training modules by the AAET, as well as by the members of the FAO-supported Networking on Environmental Education and Training (EET Networking).

The initiative to develop generic modular training materials on environmental education (EETM) to train agricultural extension workers is very useful in promoting sustainable agricultural and rural development. The non-availability of ready-to-use training materials is a major problem which the availability of EETM has helped to solve. As a training package, the EETM provides handy training materials and gives the flexibility Master Trainers require to develop and modify the content and the format of the module according to the specific needs of the learners. Preparing modular training materials necessitates cooperation between various related institutions. This cooperation is essential not only for content validation but, more importantly, to gain support from the respective institutions in the implementation of the module. The successful implementation of the EETM has made it easy for the AAET to obtain funds to develop other related training modules.

Developing a well-packaged and attractive training module calls for special module-writing expertise. Therefore, the institutions/agencies concerned must prepare their staff to develop competence in module writing. In institutionalizing the EETM, the selection of a strategic institution is a crucial factor to guarantee that the module is used and adopted by the institutions concerned. In this case, the assignment of AAET as the executing agency to develop the EETM is strategic since AAET has a direct command line to the relevant institutions, i.e. Agricultural In-Service Training Center (AISTCs), Agricultural High Schools for Development (AHSDs), and Colleges of Agricultural Extension (CAEs). The institutionalization process was further promoted by a circular from the Director General of AAET which instructed the Heads of AISTCs to use the EETM and conduct a regular evaluation on its implementation.

Dire Straits

Smoldering heaps of garbage in a huge "sanitary landfill" that isn't sanitary at all. Methane gas and dioxin smoke in your eyes, an unbearable smell in your nose, bacteria and viruses in your breath and screaming noises from bulldozers in your ears. And yet - amidst these most inhospitable conditions there are makeshift shacks from cardboard and plastic sheets, indicators that people must be living here. People in dire straits, hundreds of them, trying to survive by picking reusable material from the garbage into bags and baskets: plastics, paper, tin cans, wood. The bulldozers of the city sanitation office don't take any notice of them when maneuvering around; accidents are daily routine. Nobody talks to these people as they  *are considered "untouchables". The noise, the heat, the smell and the tears virtually envelop the waste pickers in a "culture of silence" characterized by discrimination, isolation, insecurity, fear, mistrust and a lack of communication with the rest of society.*

Rubbish, Recycling and the Role of Media - Public Awareness Strategy to a Waste Picker Development and Recycling Project in Indonesia

Manfred Oepen

This case study highlights the communication component of a project which went through two major phases that could be labeled "poverty alleviation" from 1991 to 1993 and "solid waste management" from 1994 to 1999. It is about to enter a third one, "urban development". In the beginning, the focus clearly was on the waste pickers' structural poverty, while environmental concerns were but secondary ones. This changed after 1994, when it became evident that the pickers' dire straits could only be alleviated in a lasting way if what they did for a living, namely recycling, was firmly grounded in municipal solid waste management (SWM) and was accepted by society. Consequently, more actors added to the complexity of a picture which called for local, yet concerted solutions. The local government, particularly the sanitation department, community neighborhoods and the private sector took over active roles in the project's second phase, cooperating with the pickers and the NGO that had started it. While at the initial stage, media were utilized not to "give" the waste pickers a voice but to make it heard and understood, the later stage required a more comprehensive environmental communication strategy. The strategy put the project management in a position of instigating, motivating and mobilizing for the cooperation and civil dialogue necessary to integrate the waste pickers and their recycling efforts.

Poverty Alleviation, 1991-1993

The waste pickers' problems lie in their insecure legal and low social status, and their stagnant productivity and economic dependency, which make them easy targets for harassment, eviction, corruption and exploitation from middle-men, the private sector and local authorities. The lack of access to local decision-making, loans, education, public services or the media result from this situation. Hence, the 1991-1993 orientation phase of the "Waste Picker Development Project" - an Indonesian-German development cooperation project of the Deutsche Gesellschaft für Technische Zusammenarbeit (GTZ) GmbH and the Indonesian Home Affairs Department, implemented by NGOs and academic institutions in Jakarta, Bandung and Surabaya - was geared towards structural poverty alleviation and human development by

- lobbying for policy changes affecting their legal status,
- improving their public image and social status,
- increasing their productivity and value-added of recycled products by increasing their bargaining power,
- fostering their participation in local decision-making,
- developing appropriate technologies within the context of an urban Integrated Resource Recovery System.

The main focus was on the integrated media support to the project in question which had four-fold orientation:

1 raise public awareness about the positive role of waste pickers,
2 foster the pickers' confidence in negotiating and communicating with other groups,
3 communication support for community development, educational and other activities,
4 raise environmental consciousness related to waste management, especially concerning recovering, recycling and reusing valuable organic and non-organic components.

The integrated media approach was mainly based on a combination of street theater and video, with theater being used *by* the waste pickers themselves and video in close cooperation *with* them. "Integrated approach" means that

- no media stands on its own but interacts with others,
- all media support project activities such as legal action, business development, vocational training and community building,
- action not only takes place at the local community level but also on the regional and national level of public relations and decision-making.

"By and with the waste pickers" reflects the fundamental difference between information and communication. *Information* is the one-way transfer of a signal from a sender to a receiver, disregarding understanding and feedback and often associated with hit-and-run media campaigns. *Communication* is a two-way dialogue geared towards shared meaning, which usually implies long-term social interaction. Hence, all media employed by the mentioned Program, whether high- or low-tech, did not inform *about* or monologue *to* the scavengers as objects, but put them in control of the message and the media in planning, research, evaluation and, as far as possible, in production. In that sense, the process of communication was at least as important as the finished media product, e.g. a theater play or a TV feature, as it was the former that helped to develop the scavengers' identity, solidarity and self-confidence.

The program goals outlined were pursued at various policy and administrative levels and were understood as a long-term process, not a short-lived campaign:

Street Theater and Video "Hand in Hand": Most of the work and the early stages of the overall media strategy were achieved at the *local level* by the waste picker groups themselves. The active medium employed was people's or street theater, while video played a more passive role. "Active" in this context means that theater took the lead in the media approach as it could, after some training, be managed by the waste pickers themselves at all stages. Theatre can, at any time, at any place and almost without any cost, be staged once the principal techniques and concepts have been acquired. It is most powerful when linked to lobbying and development action at the *local level.* The waste pickers' structural poverty is connected with a lack of "bargaining power". In communication terms, this means the ability to articulate the key factors of daily life for active participation in the social, political and economic sphere.

In Jakarta, the training necessary to achieve this "communicative competence" was organized by community media activists from a local NGO. The story lines not only focused on problems but also on solutions suggested and assistance needed by the waste picker groups. The latter carried out the research for the scripts on their own, for instance through family life stories. Humor and word games, local idioms and forms of interaction became outlets for criticism and eye-opening insights in a form socially acceptable to an audience of neighbors and local officials from the communities where the waste pickers lived and worked. Often, theatre performances triggered off a more continuous dialogue with and greater respect for the waste pickers. And as it is so cheap and easy to organize, it remained an asset to them even after the trainers had left. Thus full self-reliance was achieved, which at least in this particular case, could not be done with video as the waste pickers did not have an organization to take care of maintenance and management regarding this more sophisticated technology. Also, professional high-band video had to be used because a TV series had been planned. That was why video was assigned a "passive" role. However, this did not mean it was imposed on the pickers: it was always discussed with them what was filmed, how it was edited and finished and what it would be used for. The footage was edited in preparation for a pilot film for the TV series mentioned and various short video clips were used for lobbying with officials, as "starters" to meetings of waste pickers, in exposure workshops for journalists etc.

Whenever such initiatives required lobbying or legal and administrative support at a higher level, video documentaries of the theater performances or extra video clips with a special focus were often used to start the meetings before "hard facts" were presented in reports, graphs and other easy-to-assess illustrations. Video turned the "culture of silence" into social visibility on a larger scale. Usually, the waste pickers were met by their official and business counterparts with quite some astonishment and recognition after the latter had seen for themselves the articulate identification of problems and possible steps towards solutions expressed through theater and video. This was most obvious when one theater group of waste pickers performed in "Ancol" in North Jakarta, a type of Disneyland catering for the middle class, only meters away from the waste pickers' flood-prone homes ("Bintang Mas"). The theater play very much hit the point of a hypocritical consumer society that is in danger of drowning in its garbage but denies respect to the waste pickers who recycle great proportions of the rubbish they throw away carelessly. The general audience as well as the media attending the performance were stunned by the pickers' outspokenness, self-help potential and drive for social recognition.

Exposure Programs for Journalists: Radio and press journalists from *major regions* were invited to one-day workshops, moderated by a senior colleague, where "hard facts" on the waste pickers were visualized first. Then, the journalists were exposed to the "real life" of dump site shacks, waste processing workshops and self-initiated waste picker coops and schools. Very often, first-hand experience and discussions with the waste pickers resulted in more positive articles and features about their role in society.

Political Dialogue: Also, on the *regional level* exposure programs, seminars and political dialogue were held with officials, private business, banks and donors in which the waste pickers - partially using media produced in cooperation with them - lobbied for a better legal, economic and social status. The most prominent and lasting success of this dialogue was the integration of recycling data in the criteria catalogue for the yearly "Adipura Award", a much valued prize for the cleanest cities in Indonesia. Since 1993, the Adipura screening teams have asked municipalities about the "number of waste pickers", "volume of organic waste composted" or "tons of paper recycled".

National TV Series: On the *national level,* the new Educational Channel (TPI) was willing to broadcast a series of episodes on the waste pickers' living and work conditions and their ecological and economic contribution to society. Two pilot films were produced in Jakarta and Bandung, and one of them was shown in 1992. The partly documentary, partly dramatized episodes were expected to benefit considerably from the waste pickers newly established communication competence and social skills. Commentators in the series used the scenario for waste education by appeals to help waste pickers recycle waste for ecological and economic reasons. Unfortunately, TPI ran into general financial difficulties after 1992 so that no more episodes were produced or broadcast.

Award-winning Film: A German film team, closely cooperating with the project, produced a documentary on the life of waste pickers in Jakarta and Surabaya and their work in recycling. "Tukang Sampah" won an award at the Freiburg Environmental Film Festival in 1994 and was shown on German, Austrian and Swiss television. When TVE (Television for the Environment) made the film available to Third World broadcasting stations, it was shown in African and Asian countries.

In conclusion, the waste pickers were the focus of attention at all levels of negotiations and in all elements of the media strategy. The communication process backstopped by the mediating NGO provided this non-privileged group with access to small and large-scale media, and decision-making institutions in the political and economic sphere. Gradually, the waste pickers gained self-esteem and confidence in themselves, competence in formulating their problems and needs and, ultimately, respect and recognition from other groups of society. The creative potential, dignity and pride awakened and supported through managing and participating in media production could not be taken away from them again. The media involved in the strategy were not the usual "delivery system" used as an exercise of power and persuasion with waste pickers as passive "targets' or "user system". Instead, the informative, educative and entertaining capacity of each medium was utilized not to "give" people a voice but to make it heard and understood.

Solid Waste Management, 1994-1999

The project had a slow start into its second phase when it became evident how difficult it would be to transfer the successful, yet isolated elements of the orientation phase 1991-1993 to a level of significant impact. The pickers' dire straits could only be alleviated in a lasting and comprehensive manner if what they did for a living, namely recycling, was firmly grounded in municipal solid waste management (SWM) and was accepted by society. Hence, instead of the social and legal discrimination of the waste pickers themselves, their economic and ecological function in recycling was stressed. Instead of NGOs only as strategic actors, new alliances were sought and found with local governments and their respective sanitation departments in Bandung and Surabaya.

Now, the general problem analysis of informal sector groups engaged in recycling waste was characterized by their treatment as "tramps", obstacles to "development" and "pariah" of society, while these groups serve at least three important and underestimated functions. They absorb part of the otherwise state-covered social costs of "modernization" through self-employment in the urban informal sector as there is none in the rural or the formal sector. They also shoulder a proportion of the ecological costs of development by collecting and processing waste which otherwise the state would have to pay for in terms of waste transport and disposal. Lastly, waste pickers economically contribute highly to the efficiency of the formal sector because they provide raw material from recycled waste at a comparatively low price. Indeed, their record is impressive as recent assessments revealed that

- 8.5% of the total waste collected is recycled but composting of organic material so far is insignificant,
- the number of waste pickers involved in recycling of municipal solid waste is about 2.2 per ton of waste collected, each recovering about 45-75kg of reusable material per day,
- the recovery of paper, glass, plastic, metal, etc. can be increased from the present 7% of the total waste collected to 20%,
- the avoided cost due to volume reduction (from 8.4 % today to 36% at target level) can amount to Rp 7,200/ton (US$ 3.27 at 1994 prices), and the value of the extracted material can be increased from Rp 23,888(US$ 10.85) to Rp 46,793 (US$ 21.27) at the target level,
- at the present level of recycling, one ton of raw municipal solid waste will generate employment for about 2.2 waste pickers, 2.4 laborers, and 0.2 managers, and the number employed can be increased to 8.8 per ton at the target level of recycling.

Proceeding from such findings, the project's *General Goal* was formulated as "A contribution to the Local Agenda 21 has been made", and its *Project Purpose*: "Environmentally sound SWM, incorporating the informal recycling sector in selected cities is improved". These objectives were to be considered achieved by means of four outputs:

- The baseline planning framework for recycling as part of municipal SWM is established,
- Households, local authorities and the private sector support recycling,
- The skills and productivity of recycling workers and enterprises are improved, and
- Public awareness of environmental concerns is raised.

The following presentation is limited to activities of a related plan of operations which GTZ and the local government jointly carried out in Bandung in 1998-1999.

Bandung Public Awareness and Participation Training and Action Program

The Public Awareness Strategy related to various components of the project. The Bandung Integrated Recycling Center (BIRC) at a former dump site with a local population of 200-400, selected pilot areas where waste separation at the household level was supposed to be implemented, and environmental education provided to the general public. The key issue was improving SWM efficiency through recycling and composting, supported by a waste separation scheme. The more general idea of the strategy was (1) to bring major stakeholders together at a practice-oriented environmental communication training event, in which (2) mixed groups of stakeholders per pilot area were to be formed, who (3) would establish "Forum Komunikasi" on recycling in the pilot areas, which (4) would produce and use local media to support waste separation, recycling and composting, and (5) "cross-fertilize" with a mass media campaign at the municipal level.

The major actors in this training and action program in Bandung were positioned on three mutually supportive levels:

- counterpart organizations (sanitation department, NGOs) which later became facilitators in both the local authorities and the pilot areas, and municipal authorities considered stakeholders related to the issue of the program (public relations and environmental office),
- "intermediary institutions" such as NGOs, media, academic institutes, and
- "grassroots" representatives of community groups from both the informal

194

waste picker groups and from women's, youth or community neighborhood associations who were later to adopt the waste separation scheme in the pilot area.

The basic objectives incorporated in the training curriculum were raising communication skills, team building and inter-institutional cooperation, and improving public relations and participation. The two-week training program - one week of moderated inputs, group work and exercises in-class, a second week of practicing some of the 10 Steps of an Environmental Communication Strategy as detailed in PART 4 in the field - was based on the participatory curriculum provided by the author in PART 7. The author trained a moderator of a local NGO (Studio Driya Media or SDM) in the philosophy and the moderation and visualization methodology of the EnvCom curriculum. The NGO later took over the training of local facilitators and the coaching of community group activities in the pilot areas. During the training, two aspects were crucial: the use of MOVE –Moderation and Visualization for Participatory Group Events, and the sociocultural and functional mix of participants who were selected on the basis of future collaboration. A wide spectrum of participants learned how to use communication effectively in a field called recycling, which was all about "survival" to one person (picker), "urban development" to the next (university professor), "hygiene" to someone else (housewife) and "environmental protection" to yet another participant (NGO activist).

Immediately after the training, the participants established three "Forum Komunikasi Daur Ulang Sampah" (Communication Forum of Recycling) or FOKUS in three pilot areas in Bandung. As of late 1998, the FOKUS groups put into practice what they had newly learned during the training: the 10 steps of the EnvCom communication strategy from situation, actor and KAP analysis via media selection and message design to the production and use of local media. The major topics were waste separation, recycling and composting at the household and the district level. From the point at which the FOKUS groups started using the media they had produced, in spring 1999, their activities were "crossfertilized" with a mass media campaign at the municipal level. The major activities and achievements described below were still ongoing in spring 2000.

Public Awareness on Recycling in Three Pilot Areas
This initiative has been closely related to the waste separation and recycling schemes in the same pilot areas, and implemented by the same social groups, the FOKUS, established after the Public Awareness Training in 1998. Both activities and the Municipal PA Campaign (see below) have been coached by

Studio Driya Media (SDM), an NGO. Each of the three teams formed during the training established a Community Forum on Recycling (FOKUS) based on voluntary commitment and organized by itself. A situation and problem analysis identified the waste management problems found in their respective area. A KAP survey among 150 households per pilot area plus a workshop raised the community's awareness about the benefits of recycling and composting. Subsequently, the first steps of a waste separation and recycling scheme were implemented throughout 1999. An evaluation in mid-1999 showed that, in the beginning, the community groups were not used to the "bottom-up approach" used by the FOKUS organizers, since the "top-down approach" had been prominent for the last thirty years of the Soeharto regime. It took time to activate the groups and encourage them to convey their aspirations. Later, the groups became very active and outspoken, holding regular meetings every second Friday with sometimes more than 5o people attending.

In *Kelurahan Tegallega*, a mixed lower and middle class district in the southwest of Bandung for example, two public news bulletin boards were installed and numerous posters, T-shirts, stickers, leaflets, card games, handouts and a photo-story booklet with recycling-related messages were designed, produced and disseminated in the pilot area by the group itself. Each one of more than 50 regular FOKUSTEL members in 5 neighborhoods (RT) is responsible for "winning over" four neighbor families, trying to convince more and more people to participate in the waste separation scheme. It was here that the social marketing's crucial elements, economic, ecological and social "benefits" and "costs", were heatedly discussed. In March 1999, the local success story was visualized and presented to a group of waste management officials from various Indonesian cities who participated in an international training course on "Market-based instruments in Solid Waste Management". The FOKUSTEL efforts were later analyzed by the trainees as a case in point of public participation in civil society dialogue, decision-making and action planning.

In *Kelurahan Merdeka*, a middle class inner-city area, the women's and the youth organizations (Karang Taruna and PKK) were most successful in sustaining the initial drive. Local officials started supporting the FOKUS groups only after they produced tangible, direct results such as the installation of containers for waste separation in the area. As many households have gardens and as high KAP results on environmental concerns are not uncommon among the comparatively well-educated residents, eight household composters were installed. But despite these rather favorable conditions, the community at large still responds best to the FOKUS appeals for recycling if there are economic incentives such as rewards in competitions. The FOKUS group regularly stimulates this process by means of a public news bulletin board and a number of posters, stickers, leaflets, and handouts in order to spread the recycling messages in the area. A photo novella on the group's success story concerning waste separation and recycling was produced and distributed among the FOKUS members and households in the vicinity which do not yet participate in the recycling scheme. Moreover, a video film on related efforts of the group in seven after Independence Day in August 1999 was edited and shown on various occasions in order to promote and extend the waste separation and recycling scheme.

The FOKUS in *Desa Karang Pamulang*, a poor area near the final dump site where the Recycling Center is to be installed, got off to a slow start as residents in this area can only be motivated to separate waste and recycle if there are economic incentives. The latter could be pointed out during a study tour to a plastic recycling site. 65 people participated in a subsequent workshop. This group also produced and disseminated a number of posters, stickers, leaflets, and handouts with recycling-related messages. All groups also joined the municipal media productions related to the newsletter 'Bersih' and the radio programs of FORPIM (see below).

The relative success of the FOKUS groups can clearly be deduced from a comparison of a general Bandung KAP survey and the KAP results in the respective pilot areas. Here, by September 1999 all residents knew about the FOKUS groups, and almost 9o% were familiar with their objectives and at least one of their activities. More than 8o% had at least once joined debates on waste management and recycling and, hence, their understanding of related issues was well above average: On average, more than 65% (compared to 25% for the overall Bandung sample) of the respondents in the pilot areas knew what "hazardous waste" was, and more than 8o% (4o% overall) understood the function of a transfer station (TPS) and TPA. More than 75% (25% overall) knew that the organic components could be turned into compost, and more than 2o% separated wet and dry waste regularly (1o% of the overall Bandung sample did it occasionally). Only a few people (<15% in comparison to >55% overall) did not know why and for what they had to pay a waste collection fee at the RT/RW level and a waste management flat rate to the municipal sanitation company (PDK).

Despite the relative success of the FOKUS public awareness-raising, the one social group which was the original focus of the overall project - the waste pickers - is still in danger of being left behind, partly as a result of the severe impacts of the economic crisis that has affected Indonesia since 1998. The fact that as yet no formal arrangement has been worked out between the local recycling wholesaler (bandar), PDK and the households willing to separate waste, contributes to this trend. If households separate recyclable material and deliver it to the transfer station (TPS) or directly to waste traders, they will expect a small collection fee - which otherwise the waste pickers (pemulung) would earn. This is why the pickers in Tegallega - who also participated in the 1998 training and are FOKUS members as well - are afraid that their economic and social interests will be overlooked by the community organizing efforts. The NGO that facilitates the FOKUS processes is experienced in working with non-privileged social groups and is considering a problem-solving strategy directly geared towards this group of the urban poor in the informal sector which is very hard to reach. Initiatives such as skills and human resource development for waste pickers are planned in the context of the emerging Recycling Center - BIRC (e.g. recycled paper production, composting cooperatives etc.) for which a public-private partnership between a private investor and PDK, the municipal sanitation company has been established.

Municipal Mass Media Campaign on Recycling

A predominantly non-commercial approach via-à-vis the mass media was applied, i.e. instead of paying for media coverage, long-term, regular and strategic alliances and mutual support systems were established in early 1998. The general objective of the ongoing municipal public awareness strategy is social marketing in relation with the SWM services and the recycling efforts of and in the Bandung municipality. The public should understand better what PDK and other actors do where, how and why, and the information provided should instigate interest and motivate active participation in existing recycling schemes. In the process, the municipal campaign should support and complement the local public awareness activities by the FOKUS groups in three pilot areas. A monthly newsletter should provide a corporate identity and information exchange platform for the various actors involved in the public awareness strategy at the municipal and/or local level (insiders), and an outlet that can reach other interested parties (outsiders). A close and regular liaison with key actors such as journalists, schools, hotels, major enterprises, banks and the expatriate community is maintained at all times.

The public awareness activities at the municipal level focus on a "Recycling Media Campaign" conceptualized and supervised by Driya Media in cooperation with the author as an international consultant and a local campaigner. In 1999, the campaign started under framework conditions which were far from optimal: a local government unable to define the parameters and priorities of urban development in the decentralization era, soaring prices in the media sector due to the economic crisis, and a ridiculously low budget. Driya Media managed to build upon a network of socially and environmentally concerned journalists, students and NGO activists. Coordinated by an experienced AIDS and social marketing campaigner and coached by Driya Media, radio and print productions have been delivered to selected target audiences through a highly cost-effective distribution network since early 1999. The first phase of this particular program was focussed on training and network development. In the second phase, in late 1999, the program started harvesting the results from the previous training and networking activities. Now, not only professional radio and print journalists and teachers contribute to the media productions but also FOKUS group members, students and other "regular people" who follow the "golden rule" of public relations: "Do good (in this case in relation with waste separation and recycling) and talk about it (in this case through the newsletter, the radio programs and the school activities)!" This highly participatory approach is unique in all of Indonesia and promises a high degree of social and skills sustainability when the project concludes in December 2000. Some of the regular subactivities that were implemented by the end of 1999 are highlighted below:

■ **School Activists Workshops** Two two-day workshops and several meetings for 38 senior high school students and six teachers from six schools were held in late 1999. The students and teachers learned how to produce written and graphic contributions to the Campaign's motto, how to use discussion rounds and focus groups for public awareness-raising, and how to manage simple techniques for composting, waste separation and recycling. They actively contributed to the radio programs and No. 6/99 of "Bersih".

■ **FOKUS and PDK Staff Radio Workshop** A two-day workshop for staff of PDK public relations and waste consultation units and FOKUS members with 15 participants was held in late 1999 in cooperation with six staff from two private radio stations. The participants learned how to produce written contributions to the Campaign's motto and how to use discussion rounds and focus groups for public awareness-raising. As multipliers, they applied their newly acquired skills with radio producers and with school, student and community groups throughout Bandung.

■ **Radio Broadcasting** The schedule for call-in talk shows and focus group discussions as a regular dialogue between radio listeners and PDK, the sanitation department, regarding SWM services and problems in general, and recycling and composting in particular, was as follows:

- Radio Mara FM (1o6.85), every Wednesday 18-2o p.m.
- Radio Mustika FM (1o4.2), every Monday 9-1o a.m.
- Radio Dahlia FM (1o1.6), every Sunday 1o-11 a.m.

The radio stations were strategically selected on the basis of a careful audience segmentation. Radio Mara caters for the middle-aged, middle-class, well-educated urban intellectuals, Radio Mustika for urban middle-class women, Radio Dahlia for the low-income, poorly educated urban periphery groups. All stations belong to the regional network of private radio stations PRSSNI which - for political reasons resulting from the "New Order" media law restrictions - have always cooperated widely. Hence, what is produced by the three core stations is shared by 13 others, among them the very powerful Radio Oz, the favorite among teenagers in town. PRSSNI has an estimated coverage of 630,000 listeners in and around Bandung.

The twelve additional talk shows broadcast in 1999 were related to topics such as "SWM problems in Bandung", "PDK responsibilities at the TPA", "Transparency of the SWM retribution system", "Tips to reduce waste and to recycle recovered material". Contributions came from staff of PDK, private-sector recyclers and FOKUS groups. The numerous call-in questions and comments to the programs were recorded and followed up, mostly by PDK as part of its service quality improvement efforts. As a result of the radio outreach, PDK now receives an average of 15 customer calls a day compared to four a day in the period before the radio programs started. Increasingly, calls are recorded from non-customers who live in the area around Bandung or even in Jakarta who have listened to the radio programs and were particularly interested in one of the topics (e.g. hazardous waste and waste separation). Given the high number of calls during and after the talk shows and the low number of competent staff in the PDK public relations and waste consultation units of PDK, the reaction time especially in terms of complaints is still considered too slow. This problem can only be solved by systematic and sustained human resource development and through a computerized data bank in these units. PDK itself views the call-in shows in cooperation with the radio stations as a valuable asset, as they not only contribute to a trust-building process in the long run but also serve as an instrument for public service quality improvement.

In addition, twelve public service announcements related to SWM and recycling each lasting 3o seconds were produced and continue to be shared by the three radio stations. The announcements are broadcast before and after the talk shows and are repeated at least twice a week. The scheduling is coordinated among the three radio stations to achieve daily coverage for maximum impact.

■ **Recycling Newsletter**   The Newsletter called "Bersih" (Clean) has so far been written and edited by SDM in cooperation with students and activists from several universities and student networks. Between March and July 1999, five editions of "Bersih" were produced and distributed. A total of 1,ooo copies of the newsletter were distributed through the student networks, banks, schools and in public places in order to reach other community groups cost-effectively. Despite soaring prices for paper and other resource materials, high-quality printing on recycling paper with a very attractive design could be maintained thanks to private-sector contributions. The 6th edition was published in December 1999, for the first time containing contributions almost exclusively "from the people to the people" and edited by the professional SDM team. The total circulation was increased to 1,500 copies. The response to "Bersih" continues to be very positive. SDM counted more than 120 requests for "subscriptions" from NGOs government institutions, donor agencies, and the private sector in and beyond Bandung. Sponsorships from the private sectors were maintained.

■ **Elementary School Curriculum on Recycling**
The general objective of the curriculum development and contest was to establish long-term, regular and strategic alliances with, initially, a selected number of elementary schools, teachers, children and parents as well as environment-minded NGOs and university students. The learning objective of the curriculum, developed by IKIP, the Bandung Teachers College, relates to a better understanding about SWM services and recycling efforts in general, and the role and responsibility of individuals, households and other social groups therein in particular. The learning units were based more on experiential than cognitive learning and should therefore instigate interest, hands-on practical experience and foster active participation in existing recycling schemes. The IKIP team selected six elementary schools in various parts of Bandung, and twelve teachers and parents participated in the needs assessment and user expectation profile regarding the curriculum. Teaching aids and demonstrations were prepared and the "Recycling Module" was pretested in the six selected schools in mid-1999. Through questionnaires and interviews, teachers and head of schools were asked to analyze and evaluate the curriculum. The results were

discussed in a meeting among team members, teachers, heads of schools, and relevant local government institutions.

The school contest for both the children and the teachers was carried out on May 13, 1999 in the six schools, hosted by the Bandung Mayor. It was closely linked to the curricula pretested in the same elementary schools. The awards were individual, but the non-monetary benefits (e.g. an exhibition) were based on a previous needs assessment related to a school or group. Hence, the school curriculum and contest supported and complemented the municipal and local public awareness activities by the FOKUS groups in the three pilot areas. The curriculum, comprising a workbook for students, a teacher's guide and a general orientation was handed over to the local Education Department for transfer to other schools. It will be distributed by training of teachers and students in 2ooo.

Lessons Learned

It was clearly shown that the 10 steps of the EnvCom communication strategy from situation, actor and KAP analysis via media selection and message design to the production and use of local media (see PART 4) does work, and can be managed on the basis of shared responsibility between local government institutions, NGOs and community groups.

The communication processes and media productions employed in this case all have in common that they are not "about" or "for" but "with" and "by" the key actors concerned in the EnvCom strategy. The media involved in the strategy were not a "delivery system" used as an exercise of power and persuasion with the waste pickers or community groups as passive "targets" or "user systems". Instead, each media's informative, educative and entertaining capacity was utilized not to "give" people a voice but to make it heard and understood. This principle was extended even to media, e.g. the newsletter or the radio programs, or methods such as a situation, actor or KAP analysis which most projects leave to professional outsiders. Managing communication processes and producing media through a "by the people for the people" approach, however, not only reduced costs but also increased the authenticity, credibility and effectiveness of the environmental and social messages.

Mediating institutions like NGOs and various local and mass media play a crucial role in bridging the sociocultural gap between the local government and the international development assistance organization on the one hand,

and the community and waste pickers supposed to change attitudes and practices on the other. Not only did they facilitate EnvCom training, skills improvement, community and business development, but they also pop popd concepts and ideas top-down and bottom-up between those two worlds, which often do not "talk the same language".

Environmental awareness instigated by the media was almost a "side product". Not because the program planners were not aware of it but because they started from the waste pickers perspective, one of whom put it like this "I don't know anything about 'sustainable development' or 'environmental protection', to me recycling is a matter of survival". The environment and its protection had a human face this time: Because of a human interest in the waste pickers' dire straits "environmental pollution", "hazardous waste", "recycling", "composting", "waste imports" and other terms, during the project's "poverty alleviation" phase from 1991 to 1993, hit the news of the print and the electronic mass media in Indonesia much more often than before. During the project's "solid waste management" phase from 1994 to 1999, too, there were no "pure" environmental messages of the normative "Keep your city clean!" type. Instead, the communication processes that were triggered dealt with the perspective of the people concerned, e.g. the waste-related problems they observed or the specific benefits and costs regarding recycling they discussed. In the end, it did not matter too much *how* sophisticated the media or messages employed were. It was much more important to link the key actors, to build strategic alliances and to distribute responsibility between local government, NGOs, community-based organizations and the media. Others such as student associations, the school system or the private sector contributed to this emerging network.

Sometimes, it may not be clear whether certain environmental successes - e.g. the Recycling Center and the hazardous waste collection points as the first of their kind in all of Indonesia, or the recognition of the waste pickers, low-cost composting and waste separation and recycling schemes - can directly be attributed to the EnvCom strategy. Nevertheless, the constant cycle of communicating what was being done and doing what was being communicated definitely contributed to these successes.

Of course, the project did not solve all problems. The EnvCom strategy as part of the Recycling Project will be over one day and the waste pickers and the community groups will not be able to maintain all activities on their own. Some elements and results of the integrated approach, however, will last. Low-cost group media such as street theater or photo stories can be staged once the principal techniques and concepts have been acquired. Especially within a community setting, they will be powerful tools if linked to lobbying or environmental action. The experience and the sensitivity for the people's creative potentials, dignity and pride awakened and supported by managing and participating in media production cannot be taken away from them again. At the same time, the integrated EnvCom strategy opened up a wide range of capacity-development opportunities for all actors involved: Local government, especially the sanitation department, NGOs, media, community-based organizations such as the "Forum Komunikasi" or the waste picker groups.

The project management was aware of the importance of completing the "last-mile" tasks of consolidating and transferring the products, methods and lessons learned of EnvCom activities. That is why the related communication and decision processes have been documented. Also, user-friendly training modules and other transfer mechanisms such as well-organized study tours will be developed in 2000. These efforts are especially aimed at municipal policy and decision-makers so that the EnvCom experiences may be further improved, expanded and replicated in other Indonesian cities or in other fields.

The communication processes actively involving especially non-privileged members of civil society provided them with – often for the first time - access to small and large-scale media and decision-making institutions in the political and economic sphere. In the line of action, waste pickers and community groups gained self-esteem and confidence in themselves, communication competence in formulating their problems and needs and, ultimately, respect and recognition from other groups of society. When this trend coincided with the era of "reformasi" in the wake of Soeharto's fall in 1998, the "Forum Komunikasi" became "schools of democracy" for good governance and public participation in discussions and decision-making. There is hope that the established "FOKUS" offer a sustained chance for pluralistic bottom-up planning in urban and environmental management at a time of crisis in Indonesia when civil society demands a voice in those processes. Ultimately, this civil society may end the dire straits of the waste pickers, realizing that its human values and degree of civilization should be evaluated against the situation of its weakest members.

Environmental Education and Communication in Africa

Monique Trudel

The Sahel program of the World Conservation Union - IUCN addresses itself to the fundamental problems of environmental degradation in some of the poorest countries in the world. In order to reverse the destructive processes inherent in meeting urgent short-term needs, local populations must be fully involved in the management of their own natural resources. This approach is crucial because Sahelians depend on these resources to a far greater extent than do the inhabitants of richer countries.

Education is, therefore, an essential tool for building the human resource base required to counter practices detrimental to the environment. The education system in Africa, however, has not provided for a sound environmental management. Most of the methods of education were developed outside Africa and did not reflect a dynamic dialogue, drawing fully on traditional and modern knowledge. Aware of the constraints of the educational system and convinced that the rich cultural heritage of the region provides key means of expression and communication, IUCN embarked upon an Environmental Education and Communication Program. Collaboration was sought with governmental and other institutions within the region, thus leading to the WALIA project.

The WALIA Project

The general aim of the project was to encourage the development of appropriate attitudes and practices in natural resource management. The WALIA project started in Mali, introducing extracurricular environmental education in schools. Specific objectives of the program included:

1. arousing children's curiosity towards specific environmental problems and their affinity with the environment,
2. providing them with means to improve their knowledge on environmental management
3. making sure that the children spread concern for the environment and become active in their communities.

Activities were primarily targeted at schoolchildren varying in age from 13 to 18, because young people account for 45% of the population. The schoolchildren were perceived as intermediaries, and it was estimated that the message would indirectly reach eight people per child, including functionally illiterate people.

When the WALIA magazine has been distributed (see below), all schools taking part in the program receive a follow-up visit from the publishing team to develop the themes and challenge students to speak up and to think about solutions. Students interview key people in the community to collect information, and thus learn more about traditions, tales, legends, environmental management in previous times, and past and present rural practices. Once this information is collected it is possible to start discussion on potential actions. Maps and illustrations on relevant themes are also used to engage students in discussions. These follow-up sessions take place outside the school schedule or during biology classes, and students attend of their own free will. It has been observed that young people taking part regularly in the environmental education program not only become more involved in the community activities, but are also more successful in school.

In youth clubs, schoolchildren as well as other village youth could address certain issues relevant to the community in the program. With the help of teachers and by consulting the elderly, the pupils became environmental agents for their communities. Surveys were carried out to evaluate whether the message concerning the need of care for the environment was successfully conveyed and whether it brought about behavioral changes. For example, WALIA clubs provide young, motivated readers of the magazine with the opportunity to "do something". Teachers help them to find local partners for a practical activity. Young people are explained the importance of filtering drinking water and thus helped to eradicate guinea worm disease (dracunculiasis) in eight villages in the Dogon country of Mali after a campaign. The traditional authorities supported them.

WALIA in a Context - The Inner Delta of the River Niger in Mali

Environmental education in the countries of Western Africa neighboring the Sahel must link up first with traditional knowledge and community needs, and develop from there an understanding of how to deal with problems. School-children are agents of change in communities and the principal targets for education activities. In the Sahel, young people constitute up to 45% of the population. Although school attendance rates are low, students are the ones liable to pass on information, make people aware of environmental issues and the importance of resource management.

IUCN has been working in the region for more than 15 years, focussing on fundamental problems of environmental degradation in some of the poorest countries in the world. There are no simple solutions. Education about the environment, in the form of a dynamic dialogue, drawing fully on traditional and modern knowledge, is therefore of primary importance. Environmental education in Western Africa has become a priority not only for IUCN but also for many other organizations. A network has been formed to reinforce their action through exchanges and sharing of experience. This has led to the creation of a large environmental education Sahel network, operating in at least nine countries. The network was established in 1993, aiming at making environmental education an essential instrument of sustainable development. The IUCN program has proven to be a genuine way of approaching schoolchildren, making them aware of environmental problems, providing them with the technical knowledge they need for better management of natural resources, and encouraging them to take practical action for the protection of their environment.

The programs

- serve as demonstrations for future action by governments,
- support the technical government departments which are responsible in the field for raising consciousness and for teaching adult populations in the various development sectors,
- serve as a source of experience for bilateral and multilateral cooperation partners in developing educational sectors,
- provide models for neighboring countries facing similar environmental problems and willing to adopt the same type of program.

Short-term objectives are to arouse curiosity, encourage the sharing of information, and promote understanding and thinking so that positive action will be taken towards the environment. In a first step, a national team is set up and trained that is able to publish a magazine, and convey to schoolchildren a better understanding and knowledge about protection of the environment. In a second step, teachers and communication agents are trained to organize an education program that will raise environmental concerns. The long-term objectives are to

- have young people participating in the management of natural resources for sustainable development,
- apply the program in all parts of Western Africa where IUCN is present,
- share information and experience among the different partners in environmental education in the region in order to create a Sahelian data bank,
- develop among people in each country a sense of belonging to the Sahel as an encouragement to communication,
- train teachers to pass on their knowledge to the community.

The program evolved from a magazine on "how to get a better understanding of our environment", which was provided to 25 target schools as a first step. One of the key strengths of the magazine is its style. Illustrations, regular headings, boxes and simple and interactive language - the stork "Walia" talks to the children - promoted its accessibility. Ample attention is paid to suggestions of pupils and teachers. Subsequent issues are always devoted to a single major topic. The magazine is carefully and logically planned:

- subjects reflect the immediate surroundings of readers,
- each issue is devoted to a single major topic - wetlands, arid zones, water, trees, etc.,
- great attention is paid to pupils' and teachers' suggestions,
- the informal editing approach involves the reader directly and creates a sense of closeness and a need to react.

WALIA's approach to EnvCom combines traditional and modern knowledge, skills and practices. The idea behind the central instrument - a magazine provided to schools - is that children play a crucial role in mobilizing their communities around environmental issues. Students are encouraged to make surveys and organize contests. Student surveys, together with letters, help to define topics to be dealt with in future issues of the magazine. The magazine is very often the only written document in a school and becomes a valued teaching aid. On the basis of the magazine, more educational materials are developed. Twice a year, the WALIA team visits 25 schools in order to make interactive presentations. During these presentations a relaxed and enjoyable atmosphere is created, offering an opportunity to escape from the usually rigid schedule of the schools. Having discussed specific issues of environment or natural resources at school, they start interacting within their communities. Village leaders and traditional media play an important role in reinforcing the message. This message, thus, becomes relevant and accessible to the whole community, whether literate or illiterate. Concrete action around natural resources management and structural involvement of young people herein is the final step in the process.

Extending the Experience and Knowledge

The experience with "WALIA" has provided the know-how necessary to establish environmental education programs in other Western and Central African countries, each with different problems and priorities:

- "Alam, Katoutou, Ngouri", the IUCN projects in **Niger** are implemented in a desert and Sahel areas. Droughts have led to over-exploited pasturelands and intensive cutting of firewood. There are changes in the organization of society and in behavior towards the environment. People meet their immediate needs without thinking about the future of natural resources.
- In **Burkina Faso**, the educational project "Zooni", was part of a natural resources management program outside two national parks. The main problems were over-exploitation of wildlife, poaching, woodcutting and the use of land for agriculture.
- "Ekoobol", in **Senegal**, was concerned not with the Sahel, but with the sub-Guinean zone, where the mangrove forest was either in the ultimate stage of deterioration or totally devastated with no possibility of regeneration. The most crucial problem was the lack of fresh water. This program had to stop because of the war conflict. The initiators started a new program in Northern Senegal, "Djoudj National Park" where biodiversity and conservation issues are of importance.
- In **Guinea Bissau**, the IUCN education program "Palmeirinha" aims to safeguard and protect coastal landscapes, develop the rational use of natural resources and discuss the creation of national park and biosphere reserves with the local population.
- In **Cameroon**, the IUCN educational program "Yem" aims to review water management in the re-inundation of a floodplain after environmental

drought problems occurred because of the construction of a dam 20 years ago. The approach will include adults, and use different tools such as dance, theater and radio.

- In **Congo**, "Ngoulou masi" aims to make adults aware of the management of a fauna reserve. The magazine became a wall newspaper as few schools and no cultural activities exist.

The number of people directly involved in the program, i.e. schoolchildren and teachers were 7,000 in Mali, 3,000 in Burkina Faso, 1,000 in Senegal, 10,000 in Guinea Bissau, and 7,000 in Niger in the first stage. Today, there are more than 10,000 in Niger and Mali, 2,400 in Congo and 5,000 in Cameroon. Regional workshops to exchange ideas and training methods are arranged, giving people time to share and adapt the approach and tools to their realities. Experience has shown that it is possible in environmental communication and education to work in collaboration with associations concerned with the social and cultural heritage.

There are some other examples which are more indirectly linked with the WALIA experience:

Nature Clubs
The idea of nature clubs was born out of a very simple observation: in the WALIA program young readers are highly motivated and want to do something. Apart from participating by writing in, and apart from meeting the team responsible for producing the bulletin, they want to be involved in activities which enable them, together, to do something to protect their environment. Just as WALIA encourages them to do. For example, in the Ningari region in Mali where the Guinea worm is very widespread, the young people have organized themselves. With the approval of the traditional village chiefs and the help of their history and geography teacher, they have launched an information campaign. Coming together from eleven villages in the region, teams of five pupils patiently explain how you can get Guinea worm, how to avoid it by filtering water and with what. The WALIA team comes up with technical information, helps them make the link with traditional and modern authorities and convince them that thanks to clubs they can do something concrete which benefits all without financial support. In some other clubs, they took the village festivities as an opportunity to perform a ballet about locusts and a short play about UNICEF's extended vaccination program. And at Sevare, the club wrote a play in the form of songs to tell the WALIA story. These clubs need to be able

to perform ideas and require technical support and recognition from adults and projects leaders. The Walia team can help them by encouraging them to go step-by-step and by opening doors for the implementation of their projects.

Marionettes

Using marionettes and puppets to educate children and share knowledge and actions with adults creates a link between traditional and modern communication on management of natural resources. Experience has been gained in Mali and Niger in using traditional marionettes and puppets to talk about environmental issues. In Niger, having a national museum based in a capital with a large open-air area offered the opportunity to test whether an urban population would understand traditional media. In order to make this event possible, a partnership was established between the Ministry of Youth and Culture, the National Museum director, and the Police Department. The media were informed, and advertising was made possible through the newspaper and radio. The show presented two kinds of marionettes, the "big ones" representing the extinct and mythical animals and the "small ones" in cages representing the relationship between animals and human activities in rural areas. This show was based on the traditional belief about the importance of the animal in the "bambara" society and the changes that development brought about. When this show was presented in Niamey National Museum, visiting children and teachers were quite interested and many questions were asked about the media itself. The message was understood through movement, songs and dance only: "We don't need to understand the word to be able to understand the message, the visual dimension speaks for itself". The museum director requested more performances by the group and had ideas about issues for shows. The local population was very impressed and pleased to notice that their traditional media, neglected for so many years, were brought back and put to good use, as they helped them to be aware about issues such as the importance of the fauna, stopping poaching and extending land for agriculture.

After this experience, a traditional marionette festival was held in Zinder and brought groups from Togo, Niger and Mali together. In Mali, the "Tiori Ble Diarra group" was asked to stage a performance on environmental issues based on the WALIA objectives and issues discussed in the magazine. The language and writing was left to them. The WALIA team gave them the magazines and asked them to produce the "WALIA" stork puppet and use it in line with the traditional belief, i.e. as a "harbinger of nature". A show was produced and a tour was organized through the entire region, playing for adults and schoolchildren. The impact was very strong so that the troupe was asked to act as animators for a workshop, transferring their knowledge to youth in order to help them produce their own puppets and shows.

Traditional Theater

The Environmental Education component of the Lake Tanganyika Biodiversity Project of GEF/UNDP in Zambia chose a traditional media tool, drama group theater, to make fishermen and villagers aware of the issues of destructive fishing techniques. An effort was made to work with the traditional drama groups as the effective proven tool in that area. Firstly, a workshop was held to help them develop a play on fishing issues. Secondly, the drama group visited all the lakeshore villages, performing and discussing the issue with the villagers. As a result, the traditional chief decided to ban fishing with the mosquito net technique.

Theater is a very effective medium that combines humor and entertainment with reflection and message. In Burkina Faso as in Mali, theater became a tool for development. Theater groups have chosen to work in some cases in partnership with NGOs or projects and be their educational media. The play written takes in account the tradition, culture and social aspects. Plays are related to environmental and development issues, the synopsis exposing extreme responses to the issue. The public is asked to come on stage, play their role and show how they deal with the issue and how they think it should be solved. This is a very effective way of involving the population. The troupe usually works on an issue recognized as a priority by the population, and on important issues for the NGO and the local authorities. The play is written at the request of the NGO or the project, and completed after visits and discussions with the population. Finally, the play is edited and the drama group returns to the village for the presentation. The same play will be presented elsewhere taking in account the reality, the culture and the involvement of different target groups. A lot of improvisation and flexibility will be needed to treat the same issue for different situations. The objective is to have the villagers identify solutions by themselves. This will only be possible if they feel the problems presented are indeed important and relevant to them.

Games

The purpose of environmental games is to have fun while at the same time participants learn about a way of life (herders, fishermen, agriculture, nomads, etc.), an environmental issue (clean water, hygiene, importance of trees) or the conservation of endangered species (see Zschocke's contribution in PART 7). The quality of the games and the success they enjoy among adults and students justify their production in boxes or in larger sizes, and if possible their wider distribution. Games are showcased, for example, in the West African IUCN'EE&C (Environmental Education and Communication) bulletins. The center page is almost always a game that can be cut out and later fixed to a large piece of white cloth that will serve as a board on which all collected game illustrations can be placed: the pond, fish, a fish eagle, water plants, etc. To attach the paper illustrations to the cotton board, a mixture of wood shaving and glue is pasted onto the back of the drawings, a good substitute for tape. In some cases safety pins are used. The series of images that serve as the interactive element must be as large as possible. It should make the discussion about the game flexible and adaptable to the reality of the public and the issue discussed. In Mali, for example, teachers copied the elephant game on wood and played it with adults in a village. Technical experts used games to make herdsmen aware about the importance of vaccination.

The Conkouati Reserve's Environmental Communication Program decided to produce games for adults. The games deal with environmental issues such as sea turtle protection, hunting and poaching in the reserve. These games will be produced on the basis of chest games or "snakes and ladders" and will stay at the place in the village where people can meet, play and discuss the issues. They should start at the local level when local media or local awareness projects and events bring individuals, families and organized groups together who may interact directly.

Conclusions and Recommendations

Sustainable development programs risk failure if they do not take in account the experience of the people they intend to benefit. It is necessary to listen to the people to understand their behavior towards natural resources before any action is taken. The importance of environmental education and communication is not to impose a way of thinking, but to clarify and deepen the people's understanding of the environment in which they live and of the problems they face. A climate should be created in which they will find appropriate solutions through changes of attitudes and behavior, and through the application of simple techniques and practices.

The lessons learned from WALIA's environmental education and communication program are to

- listen to and understand the knowledge, experience and priorities of the people concerned,
- involve all partners from bottom to top from the beginning and throughout the whole process,
- take into account the traditional ways of communication, of passing on knowledge, skills and cultural heritage, and
- make EnvCom become a program of the people concerned, incorporating their new way of thinking and behaving, their means of transferring "old" and "new" knowledge and technology into their own words and actions.

Environmental Communication and education in Support of Sustainable Regional Development in Ecuador

Marco A. Encalada

In 1995, the Provincial Council of Ecuador's Pichincha Province (regional government) requested the OIKOS Corporation (an NGO) to develop the "Environmental Communication and Education Plan" (ECEP), as part of its "Western Pichincha Regional Development Program" (WPRDP). This program has several components. It was decided that environmental education and communication (EEC) should focus on the "Environmental Forestry" component, although they could also be expected to influence the other activities, namely health, education, agro-industrial production and municipal institution building.

The central objective of EEC was to create adequate levels of environmental awareness, enabling the people to develop a sense of responsibility in identifying solutions to local environmental problems. They were also to be partially integrated in the overall program, which seeks to secure sustainable development in an area of more than 9,000 sq km, occupied by about 350,000 inhabitants, 70% of which is rural. The plan put heavy emphasis on systematic planning and programming of educational communication, as well as on intensive training of the work team, while the overall approach was to be reasonably participatory.

It was decided that the project would have a maximum term of three years to correspond to the maximum duration of the overall program already initiated three years previously. In the first stage of ECEP, the plan was to be elaborated within three months and put into action during the remaining nine months of the year. In the second stage covering the following two years, its application would be continued and the results evaluated. This case study refers only to the first stage, and highlights the methodological approaches and the main communication planning process carried out. The content and instruments used in the communication campaigns are not described here.

The Planning Process

Three major planning steps were established at the outset: 1) diagnosis of communication and education needs in the region covered by the WPRDP, 2) formulation of concrete objectives, and 3) elaborating sets of strategy trees for EEC.

Diagnosis of Environmental Education and Communication needs

The diagnosis helped clarify: a) the state of the chief environmental problems and the natural resources of the area, b) the levels of people's environmental perceptions and consciousness, c) currently common environmental practices with respect to natural resources management, and d) the actual implicit and explicit needs of EE and EC.

State of environmental problems

This chapter of the diagnosis comprised a technical description of the environmental problems, encompassing causes, effects, and the social factors influencing them. The research on the state of the natural resources and environment sought to clarify the availability of natural resources, the state of the natural environment, the major sources of environmental pollution and natural resources depletion, the main types of social pressure being exerted on natural resources, and the main socio-environmental conflicts. It was found that the most pressing problems of the region were water pollution, soil erosion and degradation, deforestation, extinction of wild animal and plant species, and air pollution.

As for the impacts of environmental detriment, the study explored those which affect public health, the replicability of natural resources, micro and macro-economics, biological diversity, the overall local cultural development, and the landscape. The main social factors identified were associated with lack of and/or inappropriate national and regional policies, research, technology transfer, organizational development, municipal institution building, legal and technical environmental regulations, financing of environmental programs, and general education.

The main research methods and instruments used were: focus groups applied with officials and technicians from the WPRDP and non-governmental and public development organizations, direct observation in the field, and formal surveys of specimen rural and urban communities. The results formed the basis for the general orientation to be taken by the communication intervention when the ECEP was to be applied.

Analysis of environmental consciousness

This task aimed to identify: 1) The people's awareness of environmental problems, their causes and effects and the social factors influencing them, as well as of the feasible solutions available, 2) the attitudes towards confronting solutions and control of social factors that impact on the problems, and 3) the most common open behavior with respect to utilization of natural resources and the care of the environment. An in-depth survey was performed on a sample of 300 households.

Environmental practices and habits
A survey of 50 households selected by occupation was conducted to explore the main environmental practices and habits of the people. A complementary series of in-depth interviews with ten small-scale entrepreneurs was performed over a full day to gain information on the positive and negative practices employed in the various production processes in the region. It was important to determine the conditions under which people would be willing to change their habits in order to achieve cleaner production and reduce pollution.

The environmental communication and education needs
Some basic attitudes, behavior patterns and practices of the population concerning environmental communication and education were also explored. Among these were the level of application of EE and EC in schools and communities, the level of people's ability to communicate within and outside the community, types of skills and knowledge that need to be developed to improve the utilization of natural resources, the level of people's sensitization and motivation towards community participation in environmental programs, and the most influential groups within local communities. The methodologies applied were: focus groups for various segments of population by occupation, and matrix analysis of EEC needs for each community.

Formulation of Objectives

A special procedure was developed to formulate the objectives for the overall communication and education program: a prognosis mechanism to establish the minimum knowledge quantum, attitudes and practices that people would have to acquire if certain scenarios were to be realized by the end of the program. A series of single applications of prospective methods to outline foreseeable futures was applied to determine which parts of people's knowledge, attitudes, behavior, practices and habits would need to be reinforced, and which ones reoriented. Based on these results, sets of strategic objectives were set up to be reached during the term of the project.

Axiological objectives emerged as a first need, i.e. the need to consolidate certain ecological and environmental principles and concepts, such as sustainable development, people's participation in pollution prevention and control, the role of environmental communication and education, biodiversity, ecology, environment, and so on. Strategic objectives were established around the action processes that must be organized to attain certain goals, with each of the vectors called: knowledge, attitudes, practices and habits. Operational objectives were defined for each of the campaigns, based on the various specific needs of the program.

Elaboration of Strategy Trees

The most salient tasks of the planning process included the design of a general environmental communication and education model, its discussion with officials from the WPRDP, and the EEC activities-designing sessions with members of the various audiences: communities of interest, municipalities, workers, households, politicians, small and medium-sized entrepreneurs, teachers and students in the basic education sector. Two sets of strategic models were developed: the "umbrella" type strategies and the "subsidiary" ones.

The first aimed to create within the audience a common fundamental background of understanding of philosophical and axiological ecological principles, as well as of environmental concepts, on which all people should reflect at least a little and be familiar with. The second set of strategies (subsidiary) was geared to assembling specific communication and education packages to help the predetermined audiences: a) raise their knowledge levels about the particular issues of concern or relevance and b) reinforce and/or reorient their environmental attitudes, behavior and day-to-day practices with respect to local problems. Specific communication and education systems were developed within each of the strategic models to cater for the particular issues, audiences and goals. Each of these systems were called "campaigns" in order to organize the planning process.

The umbrella strategies

The umbrella type strategies pursued two directions: the first aimed to promote certain pre-established axiological environmental principles and concepts, and the second, to foment basic pre-established environmental attitudes (feelings and emotions).

The first entailed the selection of six environmental principles and three sets of environmental concepts, to be disseminated through two communication campaigns: one directed towards a group of so-called social diffusers - local people that can act as multipliers of information (Campaign 1), and another targeting the general public (Campaign 2). Under the second direction, six sets of attitudes were presented to the communities by means of a persuasive mass communications campaign (Campaign 3). Each communication/education system was organized taking into account the following elements:

- The structure and functions of the system, as well as its technical components, such as the message, media, communication senders and audiences.
- Duration of the campaign
- Specific goals
- General activities and operations.

The subsidiary strategies

Under the "subsidiary" strategies two approaches were applied:

- The first approach perceives communication and education as "preconditions" for the changes required with respect to the social factors that affect the causes of environmental problems. For instance, definition and approval of environmental policies, regulations, and both funding and technology transfer programs are seen as a way of initiating the process of solving any problem associated with water pollution.
- The second approach perceives communication and education as "tools" to encourage specific behavior in people regarding proposed solutions to concrete problems. For instance, once certain environmental programs are put into action to prevent environmental damage in any production system, communication must be used as a tool to ensure that cleaner production technologies or production practices are adopted, such as better management of raw materials and wastes.

Under the approach of communication and education as preconditions, two systems were developed: to promote the approval and application of new environmental norms and regulations at regional level (Campaign 4), and to motivate decision-making in favor of complementary local environmental management programs to be launched by the municipalities and local development agencies (Campaign 5). In turn, the strategic approach of communication and education as tools resulted in the development of systems to influence the informal and formal education sectors.

The informal education system sought to create opportunities for a wide sector of the population to become well informed and able to conduct and/or achieve:

community dialogues, consensus and negotiations concerning: 1) the description and interpretation of environmental problems (Campaign 6), and 2) the day-to-day practices people should observe to prevent pollution, to protect the environment, and to conserve natural resources and wildlife (Campaign 7). Two practical ways of addressing environmental problems emerged:

- an approach to tackle various problems as a whole where the role of synergy is important when they are detailed (Subcampaign 6.1),
- a single problem approach, allowing problems to be addressed individually with their own characteristics, effects, causalities, social factors, and alternatives for solutions (Subcampaign 6.2).

The formal education system addressed the work from two perspectives: Efforts were made to help develop the environmental education technical infrastructure in the school system of the region, such as curriculum revision, teacher training, educational materials production, and teachers were motivated to perform environmental communication and education in schools.

The Implementation Process and Results

The management team in charge of the program comprised professionals from various sectors: communications (2), informal education (2), formal education (3), production engineering (4), environmental law (1). They all received intensive and ongoing training on how to interpret the campaign-designing process, how to implement the campaigns based on people's participation, and how to use feedback to reorient activities.

A parallel supervision system was mounted so as to analyze the performance of the planned activities and people's reactions to the program implementation process. The plan has been implemented in full. The preliminary results have shown outstanding participation and a good response by the people to the operational model and the working processes. The various audiences feel that they have received valuable support from the EE and EC program. Communication instruments have consistently been produced with high levels of participation and consultation by the people. Pretesting of materials and validation of them in practice have been important routine activities applied so far.

Some other reports on the performance of the project have shown that people in the communities have been highly motivated to participate in the various activities suggested by the program, and have developed an interest in modifying their production processes from the environmental perspective. Some have even already initiated changes in various sectors of production and at home. However, formal evaluation is still required and is expected to be carried out at midterm of the implementation stage.

The main constraints established over the nine months of the implementation process were the lack of institutional support from the Provincial Council in encouraging the implementation of the communities' initiatives with respect to environmental projects resulting from the communication and education program.

Environmental Education and Communication on the Black Lion Tamarins in Brazil

Suzana M. Padua

Conservation education programs for natural areas in Brazil are rare. Natural areas are not commonly utilized for education purposes and consequently people are unaware of their importance. Nevertheless, in the education projects designed for the conservation of the lion tamarins *(Leontopithecus)* this has not been the case. The first education program established for the conservation of the golden lion tamarin *(Leontopithecus rosalia)*, which began in the early 1980s, opened a new conservation scenario which was adopted later by educators who designed programs for the other lion tamarin species (Padua et al 1991). These programs may vary according to their context, but in general they are designed to involve local communities in the conservation process and to disseminate scientific findings in a simple and direct language so information can be understood by all. As the lion tamarins are charismatic, they have become symbol species to attract attention and build up pride, enhancing the protection of the natural areas where these primates are found. Additionally, some of these education projects have applied research methods which are important to improve the strategies used, to help assess their overall effectiveness, and to disseminate information on what has been effective and what has not, so other educators may benefit by avoiding the same mistakes (Dietz/Nagagata 1986, Padua/Jacobson 1993, Jacobson/Padua 1995, Padua 1993, 1995).

The environmental education program for the conservation of the black lion tamarin began from scratch in 1988/89 at the Morro do Diabo State Park, administered by the Forestry Institute of São Paulo, the most significant habitat for this species. Education initiatives at other sites where black lion tamarins are found have covered the following: a course for teachers in June of 1992 at the Caetetus Ecological Station, also administered by the Forestry Institute of São Paulo, which served as a starting point for a continuous school program carried out by the park's new management; activities with the local students included a study to assess how much information parents learn from their children at the Fazenda Rio Claro.

The education programs for the black lion tamarin have followed a systematic process of which the most thorough has been at the Morro do Diabo site, which will be briefly summarized in this paper. The Morro do Diabo State Park education program was continuously and systematically evaluated according to the PPP model (Planning/Process/Product), designed by Jacobson (1991) and adopted by Padua and Jacobson (1993) and Padua (1997). This model helps ensure effectiveness in each step of the program from conception through completion, or through the planning stage, the implementation process and the product, or summative evaluation.

Planning Stage

During the planning stage the needs, goals, objectives, target public, constraints and available resources were defined. A preliminary survey conducted among the local population showed that people had very little prior environmental knowledge. Although people showed great interest, the majority knew little about the local flora and fauna. The need for a broader knowledge and understanding became evident. The goals and objectives of the program were also defined based on information gathered through the surveys. Since the park is the last large remnant of Atlantic forest in the interior of the State of São Paulo and therefore highly threatened, the main goal was the conservation of the park itself. The objectives were to foster among local people an appreciation of the park and its rich biodiversity.

A specially designed school program introduced students to the park and furnished means for them to increase their knowledge on ecological concepts and to shift their attitudes towards nature. Strategies that would impact individuals' values and knowledge were purposely applied during all the stages of the program, for research has shown the importance of these aspects to increase awareness and change people's behavior (Lozzi 1989, Hungerford/Volk 1990). Although the environmental education program targeted mainly the local students, many activities were especially designed for broader public participation.

Involvement of the surrounding communities in the conservation of Morro do Diabo was of great importance due to the accelerated rate of the destruction of the local natural environment. Students alone may not have a chance to alter the destructive process, since little might be left for them to fight for when they are of age to become the decision-makers. Therefore, the environmental education program sponsored several outreach activities targeting all community members, from local authorities and businessmen to laborers.

The planning stage also included seeking institutional support and participation, which were crucial for the program's implementation. The park's employees were encouraged to cooperate and as a consequence, trail signs and activities were designed and mostly executed at the park with minimal extra resources. Although the Forestry Institute of São Paulo was very supportive of the education program and helped in many aspects, additional support was obtained from several institutions interested in contributing to the park's conservation. This was an important strategy, because it facilitated the program's implementation at a faster pace and could be based on the priorities set by the education staff.

Process Stage

In the planning stage the program's content, implementation strategies and evaluation procedures were defined. The content for the black lion tamarin environmental education program was selected based on the information gathered in the planning stage and in the scientific findings of a long-term study on the species (Valladares-Padua, 1993). Program strategies were designed accordingly and included the elaboration of educational materials prepared for the local teachers, who lacked information on the park, its natural resources and local history. Visitors watched a slide presentation before their visit to the park. Three nature trails each focussing on a different aspect of the park were adapted for students' visits. A visitor center included an exhibit area and an area where objects could be handled through especially designed activities to stimulate curiosity and provide learning stimuli. Students had the opportunities to play games and have live animal contact. As snakes were especially feared in the region, the education program kept three snakes in captivity, which could be handled by learners to change their feelings of fear into respect. After the visits, students received handouts with games and information, and class contests were sometimes held.

All activities were designed to encourage appreciation of the black lion tamarin and the park as a conservation unit. Each activity was pilot-tested and constantly evaluated. This process evaluation furnished helpful information to improve the program as it was being implemented. Among the many community-oriented activities were art exhibitions, art or sports competitions and workshops. The local radio played an important role in broadcasting special activities to home audiences or furnishing information on the environmental education program in general. Two park employees and local high school students were trained as nature guides. They helped design the program by contributing new ideas and activities, which were pilot-tested and - depending on the results - adopted by all as an educational strategy. As most of these nature guides were members of the local community, they helped solve specific implementation problems and made it easier for the program to be accepted by the community.

Product Stage

The product stage assessed whether the goals and objectives were achieved, as well as the direct and indirect effects of the education program. A formal evaluation procedure helped assess the program's effectiveness. The results were used to improve, change or drop program strategies. Results based on the systematic data collection also helped to obtain institutional and funding support and were used in a number of publications which aimed at disseminating the methods applied.

The black lion tamarin environmental education program aroused great public interest, acceptance and participation. By the end of the first year 6,000 students had visited the park, and the average in the following three years was 8,000. In a systematic evaluation 144 students were assigned to experimental and control groups, and tested on three occasions: the pretest (before being exposed to the program), posttest (immediately after the park visit) and retention test (one month later). The tests were written questionnaires which measured students' knowledge and attitude. Statistical analysis determined that there were significant differences between the experimental and the control groups, indicating the program's effectiveness.

Other indicators of the program's effectiveness were: the increase of families visiting the park during the weekends, university students spending weekends at the park's lodging house, local teachers requesting environmental education courses, and the nature guides' increasing interest in improving their performance. Several events showed people's interests in the park's conservation. Some events were related to festivities, such as a float for the town's anniversary, the end of the year Lion's and Rotary parties, and other public initiatives. However, the most important indications of community involvement related to the park's own protection. After a radio interview in which the education staff explained the threats to the park's flora and fauna due to the relocation of the city's garbage disposal site, people wrote to or telephoned the mayor requesting an immediate solution. The garbage was moved elsewhere in less than a week. The community also helped extinguish a forest fire together with the park's employees. Many fires had occurred before, but the local people had never helped. This may serve as an indicator that the education activities increased the awareness and motivated people to act. Another instance of community participation in nature conservation was not within the park's boundaries. A nearby farm was being illegally logged and through public pressure the logging stopped and the farmer was fined. To improve local people's socioeconomic conditions, a group was formed among the local businessmen with the purpose of setting up development plans. All the projects approved had the priorities of not polluting the environment and of absorbing the underprivileged and unskilled population, instead of importing solutions which would not have proven sustainable. Finally, the community became active when demanding the continuation of the program itself. Letters were sent to the park's administration in São Paulo and to the local mayor requesting a local director and the continuation of the education program, which had stopped during a change in leadership.

Conclusion

The environmental education programs for the lion tamarins together with other conservation measures should serve as examples of effective and integrated efforts towards species and habitats protection. Through public awareness programs, people increased their information and shifted their values and attitudes, which may have encouraged them to act. The empowerment of local people has shown that individuals can greatly contribute to the conservation of natural areas. In Brazil, this is especially important due to the richness of the natural environments and the lack of resources to protect them. Therefore, the black lion tamarins example should be shared and disseminated so other species and ecosystems may benefit from the lessons learned.

Applied Behavioral Change in a National Park in the United States
William Smith

The ABC framework outlined under Step 2 of the EnvCom Strategy in PART 4 was used in the context of an environmental protection project to encourage people to separate garbage in two public parks at the foot of the Washington and Lincoln Memorials in Washington D.C., in the interests of better recycling. The purpose of the specific social marketing exercise was to assess the behavioral characteristics of an existing program.

Four sub-areas were delineated for observation purposes:

- near a concession stand where an attractive signboard on the importance of recycling had been placed, alongside a clearly marked container for plastic, glass and aluminum plus a traditional garbage bin,
- a street corner, not served by the concession stand, but with a recycling container and a traditional bin,
- a highly congested bus stop in front of the Lincoln Memorial where the same signboard had been placed,
- a concession area near the Lincoln Memorial with multiple garbage receptacles, the recycling sign and a rest area for visitors.

Existing signs

The signboard made three points. It instructed the public that glass bottles, plastic bottles, foam cups and take-out containers, and aluminum cans should be placed in bins separate from the other garbage (the articles were depicted and there was also an illustration of two young children using the recycling container). Secondly, it illustrated the technology of recycling, and finally it indicated how much garbage is produced on the Mall every week. The recycling containers displayed the symbol of recycling plus the words: "plastic, glass and aluminum only", and two symbols indicating that hot dog and potato chip wrappers were not to be thrown into the can. The garbage bins were of a different size and shape from the recycling containers and were labeled "Trash only". There was a third container unlabeled and different from the other two, located next to a concession stand.

The hypothesis was that the presence of new signs and the availability of recycling trash containers would lead to a high level (80 per cent) of proper disposal of recyclable drink cups and other recyclable products. The three activities in the case study were to determine whether the hypothesis is accurate (behavioral assessment and direct observation), to determine which problems in the observation areas might be solved by intervention (problem analysis and data assessment), and a brainstorming session on possible interventions to correct identified problems (program planning). Observations were conducted on two days - weekend and weekday - for two periods of 60 minutes each. Teams of two to three observers worked at each site for a total of 12 or 15 hours. Observers were trained for one hour on how to observe and record data.

Results

Most people took note of the expensive, well-designed and well-placed signs and did not recycle properly. Observation on the area and the contents of the bins showed that cups clearly marked as recyclable were almost evenly distributed in the recycling and non- recycling containers. One conclusion was that people came to the Mall to have fun - not to recycle. They expected to see signs with information about the monuments they had come to visit - not about recycling. Another conclusion was that there were too many types of garbage bin and too many conflicting labels, which led those people who tried to recycle to make mistakes or to give up.

Finally, the information program was designed to reach everyone entering the area, but only two percent bought any item that could be recycled. The general approach of public signs was thus a "hit or miss" strategy that missed more often than it hit.

A suggested strategy

A marketing strategy would lead to quite a different solution:

1. A clear goal would be set: 90 percent of recyclable cups should be placed in a recycling container.
2. A target audience would be identified : all individuals purchasing recyclable cups at the concession stand.
3. Containers would be standardized and made easier to distinguish between garbage and recycling containers.
4. Signs would be replaced by a training and incentive program for sales staff at the concession stand. They would be trained to provide a simple verbal cue to every customer buying a drink in a recyclable cup to instruct, remind and

thank people for recycling that cup. Staff could be given a small financial reward every day the recycling goal was met.

Conclusion

Environmental education must give people the cognitive tool to judge what is environmentally sound. The knowledge that serves as the basis for environmental practices and policies increases rapidly as science progresses. People must be educated to assess new developments and integrate them into the broad context. At the same time, environmental change is not only possible but can be positive and beneficial for them today, as well as for the future. Development agents find increasingly complex the job of helping communities to decide how best to protect the environment and at the same time assure benefits now. The choice of a comprehensive framework for behavioral change gives these managers a more accurate idea of the perceptions of individuals and communities concerning the alternatives proposed.

Communication between Farmers and Government about Nature in the Netherlands

Cees van Woerkum and Noëlle Arts

In 1990 the Dutch Parliament accepted the Nature Policy Plan (NPP) based on a network of areas with special natural values to save and develop nature over the next 30 years. Farmers in these areas are encouraged to restrict certain farm operations (mowing, fertilizing), with compensation for any loss of productivity. They may also sell part of their land if the land is especially important for the conservation or production of nature. The decisions are voluntary. Implementing the plan was not easy. There were signs of non-acceptance.

Farmers' Perceptions

Farmers look at nature in different ways than those expressed in the NPP. Nature for them is everything that grows and lives around them, and they react negatively to the idea of "wild nature". The process of accepting an idea or policy has different dimensions. The problem must be understood as being serious, government intervention understood as inevitable, and the main lines of the policy accepted as well as the specific measures. The measures must be perceived as effective, realistic and adaptable to the farmers' practices, and fair compared to what is asked of other people. In the Netherlands farmers did not think the current condition of nature was so bad, and government intervention was not seen as necessary. The main lines of the policy - more nature, better nature - were not received with enthusiasm. Moreover, farmers have to contend with other government interventions in the area of environment and are tired of regulation.

Farmers do not see the measures as being effective and are not convinced that there is enough money to pay for what the government wants. Furthermore if a farmer's land is in a nature conservation area, they see that their land loses value and they will be disadvantaged by not being kept up-to-date with new standards in farming. In addition, farmers do not accept that they have to bear a reasonable burden to enhance nature, as they do not see themselves as spoiling nature in the first place. In most other policy interventions it is the spoiler who pays. Farmers see themselves as victims of what they have created.

Communication and Miscommunication

The communication activities organized by the government set up a chain of mistrust due to the character of the meetings. Farmers were supposed to listen to the government information, but were instead strongly motivated to speak about their own problems. The two parties often engaged in one-way communication, resulting in ever more frustration, rather than understanding. Farmers' reactions are influenced by collective opinion in their circles based on shared goals. They stereotype the government and environmentalists – the other party – based on the latter's worst possible characteristics, compared to the best characteristics of their own group. This image gives the farmers an identity and mirrors and justifies the aversions they have to outsiders. Farmers are used to the idea that they are the masters of the countryside. Stigmatization maintains power by excluding attacking outsiders, and reinforcing the cohesion of the farmers. Stigmatization blocks effective communication and has frustrated attempts to communicate about the NPP in order to create acceptance. However, individually, some farmers are more pragmatic, especially when confronted with concrete proposals to change their practices with financial incentives.

Policy-making Process

These problems stem from the usual method of policy-making, with the government setting the goal and constructing a set of instruments to implement it. However, governments tend to look at the world according to their own definitions of the problems and perspectives on solutions. Perspectives outside the policy system are usually not taken into account. In the Netherlands, powerful interest groups pressure for social change in the environment and these are the ones involved in the policy-making. There is not much consideration for the groups who will be affected. Furthermore, contacts with the agricultural community are restricted to the top-level representatives. Typically, no analysis is worked out about the process of change that farmers are expected to undergo. Data used in the analysis for policy is derived mainly from biological and ecological research. The rationale of the farmers, and how they perceive nature and nature-related policies were not dealt with in a serious way.

The problems also extend to the communication strategies based on this policy-making approach. The communication follows a DAD model: Decide, Announce and Defend. Communication is brought in after the plan is developed and although it is informative and motivational, it does not achieve the desired acceptance level. Campaigns are typical of this instrumental approach. This manner of communication neglects the cultural factor that people's rationality or perspectives derive from the group to which they belong. It overlooks the fact that people change as a result of discussion about issues that they consider important. Another pitfall in the communication process is to forget that groups of people have an autonomy in shaping their own messages. People cannot be compelled to think in another way. In many encounters, farmers were told to to accept NPP ideas based on insights, arguments and data. However, farmers could not be convinced in this way. This is because they construct their own messages using their own perceptions of nature. They label the source of information from the government as from an "outsider group", not to be taken seriously and as a result they do not learn anything about what is being said.

A different approach

In an attempt to overcome the problems, a more interactive policy approach was tried at the regional level of the Environmental Cooperative de Peel (EC De Peel) in the southern provinces, influenced by a change in thinking as expressed in the contribution on "government as communication" in PART 3. This region is characterized by rare and beautiful nature, and a very threatening type of agriculture - bioindustry, mainly pig breeding. Could interactive policy-making result in a more acceptable plan for preserving nature, and what factors would help or hinder this process? The government policy-makers have to adopt a more flexible approach to pave the way for new regional proposals and pay for initiatives that could be more effective for a better environment. The relationships between three groups of organisations were critical to this cooperative network.

1. The governmental role is very important in bringing people together and linking the regional ideas with the national policy process. A special team was installed to organize this process in the provinces.
2. The nature and environmental organizations, especially the WBP, the committee to preserve de Peel, was originally attacking farmers with lawsuits. They moved to a more cooperative position.
3. In the agricultural community, only a small proportion were willing to engage in the negotiation process in a proactive way. Most of them kept their old positions of protesting and coping. Only gradually and with a little encouragement from their formal organization did they become more open.

The process of interaction resulted in a common plan supported by all the parties, to develop the region in an ecologically and economically viable way. The plan has been accepted by the government as an experiment, not a final plan, and the results will be checked carefully. In the process, the actions of the EC De Peel, in discussions with the government, were confronted with bureacracy and slow decision-making. The government is uncertain about the role to play, as a facilitator, rather than regulator. And therefore it is unclear whether the approach will be supported in the long run. Tension stemming from the former antagonism relaxed when the WBP moved to a more cooperative position. There is now respect for the other's opinions and the motives behind them, and there is a relationship founded on basic concern for each other. However, the relationship is fragile and some old reflexes exist, leading to litigation. Not all farmers were involved in the interaction and cling to their stereotypes of the environmentalists as being dangerous strangers from the outside. The differences between the two groups are still enormous.

The EC De Peel has to balance its role as a negotiator in a situation of conflict and its role as a mediator. As a negotiator it has a tough attitude towards the other parties, and just minimum contact with its own constituency. As a mediator it has to show a friendly face outside and cultivate contacts inside to guarantee that the farmers stay involved in the process of learning and collaborating. EC De Peel had a difficult task as network manager. Not only did they have to balance the interests of all the actors in the network, they also had to harmonize the change of one actor with the change of another actor.

To create a common focus, discussions are the basic mechanism by which learning processes and negotiations are started. The discussions tend to be restricted to representatives of all the parties. The representatives have two possible roles : they can act as promoters of certain interests or act as change agents both internally towards their own group and externally towards the other party. In the former role, the tendency is to keep quiet in the negotiation process as they do not feel they have to offer anything until the process is finished. In the latter approach the role is a busy one, trying to involve everyone in the process, to ensure that everyone says what they feel is necessary and support the outcomes. Ultimately, acceptance is not based on outcome alone but also on participation in the process itself.

Behind the two roles mentioned there are different conceptions of the negotiation process:

- distributive negotiation where everyone is after a piece of the cake and
- integrative negotiation where everyone is involved in creating the cake they want.

Distributive negotiations start from fixed positions to which each wants to cling as tightly as possible. People ask for too much, knowing that they have to give something up. The negotiators are tight-lipped about their underlying motives and personal feelings. Threats are common; the constituency is kept alert with actively distributed images of the bad enemy. Integrative negotiations start from an interest or an idea about the desired future. Understanding of the issues relating to biodiversity best comes from involvement in critical reflection. People are more open and try to share their feelings, beliefs and motives. Threats are minimized, keeping the functional relationship as good as possible. Joint fact-finding is common, and there is concern about the consequences of a discussion for the other party. Most importantly, people start learning - learning to see themselves from the position of the other and over time relationships are built up. Such learning processes are absent in distributive negotiations. Interactive policy-making involves the actors in learning from each other, understanding their interdependency, and creating more effective policy plans together.

Distributive negotiation	Integrative negotiation
• Getting a share of the cake	• Creating the cake
• Starting from fixed positions	• Starting from an interest in a desired future
• Overcharging	• No overcharging
• Tight-lipped about underlying motives and feelings	• Open, sharing of feelings, motives, beliefs
• Frequent use of threats	• Minimal threats
• No joint fact-finding	• Joint fact-finding
• No concern for consequences on the other	• Concern for consequences on the other
• No relationship-building	• Relationship-building
• No learning effects	• Learning to understand the other's position

Characteristics of distributive and integrative negotiation

Lessons Learned

The following contributions try to draw conclusions and lessons from the case studies in PART 5 by comparing and taking into consideration the conceptual framework of EnvCom in PART 3 as well. The individual contributions were written by Ronny Adhikarya, Winfried Hamacher and Manfred Oepen while others were adapted from "Beyond Fences" (IUCN 1994).

Involving the People

Information Campaign

A campaign to inform people about the environmental initiative at hand, its goals, ways of working, its benefits, and the ways in which local people and groups can become involved and gain from it should be launched (see for instance Encalada, Oepen, Trudel in PART 5). Prejudices or false information about the initiative should be dealt with first. Potential costs and information about what the initiative will and will not do should be pointed out. Wherever possible, strategic alliances should be built with local institutions, schools, NGOs, women's groups, community-based organizations, government, and cultural and religious institutions. In order to account of different literacy levels among the stakeholders and to adopt suitable communication methods, at least some information tools which are not dependent on literacy should be used, e.g. community meetings, street theater or pictorial posters. As a first step, appropriate ways and means to reach specific user groups are to be investigated. Freely distributed information can help build trust between the management of the initiative and the local stakeholders. A comprehensive information campaign can also greatly increase the level of local awareness, not just about the initiative but about the general state of local resources. Such a campaign will foster a better understanding of the initiative's benefits and costs in both the long and short term. As a word of caution, problems may arise if information about the initiative is inadequate. Faulty or conflicting information can create a suspicion that the managers are "hiding something".

Public Relations Service

If the environmental initiative is large, a public relations service can be set up (see for example Encalada, Oepen, Rogers et al in PART 5). It should be a place which people can visit to ask questions and offer alternative ideas. It may also be a place to disseminate information, an entry point for relevant databases and, possibly, a coordination center for consultants and training. Even if the initiative is small, the staff should ensure that local people feel welcome at all times. An area with an information display about the initiative can be provided that shows how further information can be requested. A highly visible suggestion/complaints box will work best where people know how to write and are comfortable writing comments. If local actors are to influence these processes, they need to be aware of how they operate and of the responsibilities of the various agencies involved. Brochures and posters, presentations to schools and churches, guided tours of the environmental area, and audio-visual displays should be considered. Media and messages work best if they are produced and flavored by local people (artists, teachers, business people) in the local language, and if they are up-to-date with solid content, to be as useful as possible. A system should be put in place which ensures that all requests for information are dealt with promptly and that people are kept informed of actions taken in response to any suggestions or complaints. By collecting views and information, the service can also act as a monitoring mechanism, picking up local perceptions, identifying sensitive issues and stakeholder conflicts as well as positive experiences related to the initiative. It can also be the basis for networking on key issues.

Environmental Discussion Sessions

Discussion sessions in local communities, in the local language, should be organized emphasizing a dialogue approach and using techniques and tools that are culturally appropriate and appealing (e.g. theater, games, audiovisuals, competitions). Information on the initiative and its benefits in the local area can be included (see for example Trudel, Encalada, Padua in PART 5). There are many ways to discuss environmental matters that are user-friendly, fun and involve the whole community, including children and the elderly. For instance, helping people to develop a theater play, slide show, photo story or video on local problems and resources can be very effective in raising awareness. The need for environmental awareness should be presented in a non-judgmental way and ideally should arise spontaneously from discussions. People may not be aware of the problems created by their actions or they may be aware of the damage but have few options (e.g. an influx of migrants may have reduced the land available or modern schooling may have meant a loss of traditional knowledge). If people are struggling for survival, they may have no alternative but to rely on the resources in a protected area. Discussions allow the project management staff to learn about local people's rationale for their actions. Open-ended discussions may improve their understanding of the causes of environmental problems. Staff can then look for solutions that local people feel are beyond their control. Once people have assessed for themselves the importance of conserving natural resources, encourage them to discuss what this implies for their life and work, the costs and benefits of changes, and possible activities to limit costs and optimize benefits. When project staff contemplate a new activity to provide alternative income or replace resources, they would be wise to hold a series of these discussions as a way to sound out public opinion. Regular follow-up sessions should be held as one cannot expect a single event to have an impact. Scheduling regular sessions will be appreciated by local people as evidence of staff commitment. Also, the discussions need to be moderated as a lively and meaningful dialogue is likely to include differences of opinion.

Helping Stakeholders Organize

Where there are power differences that disadvantage some stakeholders, the balance may improve if such stakeholders organize themselves in formal or informal ways. Such stakeholders should be assisted to this end, e.g. by offering information, training in managerial and financial skills, access to credit, opportunities to meet with organized groups, opportunities to discuss issues with specific bodies, access to technical, organizational and legal advice etc. (see for instance Rikhana et al, Oepen, Trudel, van Woerkum et al in PART 5) In particular, non-organized resource users – e.g. waste pickers in a recycling project - could be assisted to represent themselves in discussions regarding the environmental initiative (travel support, daily allowance, etc.). Every stakeholder will have different information, concerns and interests which need to be considered and developed. Making sure that all stakeholders are able to develop their own position and form of representation may initially result in more challenges to the initiative. In the longer term, however, through mass mobilization or putting local knowledge to good use, the initiative can greatly increase the level of local support and provide an effective counterbalance to destructive outside forces. In providing such assistance, it is important that the approach is compatible with the culture and practices of the stakeholders concerned. Whenever appropriate, regular social gatherings should be used by adding the environmental issue to existing agendas, rather than holding separate meetings. Setting up new organizations should be avoided unless there is no alternative. The duration and scale of the initiative will affect the extent to which stakeholders need to organize. It must be remembered that building organizational skills among a disparate group is always a slow process. People need to feel that being part of an organized group is necessary to protect their interests.

Institution for Conflict Management

Local people should be asked about traditional methods of conflict management (mediation, negotiation, etc.). Again, one should build on what exists, identify a relevant new body (e.g. a local council) or nominate an individual to mediate and deal with conflicts between stakeholders and the initiative's management, or among stakeholders (see for example van Woerkum et al in PART 5). This body or person should be widely respected, and have the trust of all parties involved, particularly indigenous groups. Gender issues should be kept in mind so that both men and women have confidence in the system adopted. The mediating body must be sensitive to power imbalances between stakeholders (users, regulators, etc.) and be able to maintain a neutral position in the conflict. There are two main kinds of conflict: conflict among users, and between users and managers/regulators. Each may require a different approach. Conflict among users is often resolved by a commonly accepted mediator. Social and community pressure for compromise can also help. Conflicts with major power differences are more difficult. Often there is a strong sense of mistrust between the parties, the sides are not equal in strength, community pressure is ineffective and there is political pressure to settle issues quickly and without compromise. Several factors are particularly important. The conflict management institution must not be seen as being aligned with any party, including management. Those entering into agreements must have the authority to represent their groups. And the conflict management institution must have coercive and/or moral power to enforce agreements.

Participatory Appraisal and Planning

Participatory appraisal methods (e.g. PRA) should be facilitated by and with a variety of stakeholders (see for instance Rikhana et al, Oepen, Trudel in PART 5). The local biological and socioeconomic environment, existing interests, capacities and concerns relating to the environmental initiative should be dealt with. Specific proposals can be facilitated through participatory planning which can be submitted to the judgement of the local residents at large. This process safeguards against technocratic planning being imposed from outside, which too often ignores the interests and capacities of local communities and other stakeholders. Involving affected parties in identifying relevant issues and potential activities can increase their knowledge and appreciation of the initiative and give them a sense of ownership in its future direction. It can also help to reduce the potential for conflict in the implementation stage. In turn, involving the staff and management of the initiative in the process gives them a greater understanding of the concerns and capacities of various stakeholders. Several issues need to be considered before adopting this strategy. Firstly, the process requires the time and involvement of experienced and trained facilitators. Also, if the stakeholders do not anticipate substantial benefits, they may be unwilling to commit time and resources. Secondly, the commitment of the decision-makers to take into account the results of the participatory planning process should be ensured. Failure to do so will create frustration, disappointment and distrust among the participants. Related to this is the need to make sure that the inputs from the stakeholders are properly recorded, interpreted and utilized so all parties concerned gain the maximum benefit.

Participatory Monitoring and Evaluation

Regular participatory monitoring and evaluation to review objectives, the approach, activities and results , should be undertaken with the stakeholders, and, if appropriate, the donors (see for example Rikhana et al, Encalada in PART 5). Monitoring enables problems to be identified and solutions to be sought at an early stage. It can be carried out on a formal or semiformal basis by both local people and staff of the initiative. A system to record the results over time should be established so that the resulting data can then be part of the evaluation process. Aspects to be monitored could include effectiveness of information systems, regularity of staff visits to communities, maintenance of park boundaries, compliance with meeting schedules etc. Stakeholders could be given authority to monitor the quality of services provided by the initiative, e.g. interactions between the local community and management, follow-up to complaints, etc. If stakeholders are responsible for monitoring, there should be a process for feeding results back to management and a commitment on the latter's part to take these results into account. Failure to do so will create frustration and distrust among the stakeholders, which could hurt the initiative. Keep a record of monitoring results, including recommendations to improve the initiative's design, management and scope. Evaluation should reassess the design and objectives of the initiative and assess its impact on the environment and the affected communities. It should be conducted in open meetings with as many stakeholders as possible, and suggestions for improvement and open discussion of the pros and cons of several courses of action should be solicited. Appropriate questions include: What is getting better? What is getting worse? Who is gaining from the initiative? Who is losing?

Addressing Local Needs

Special Events and "Ideas Fairs"

Special events can be organized to elicit new ideas for initiatives to link local livelihood with environmental initiatives. Awards can be provided for the best ideas ("ideas fair") and activities, and linked with sports matches, market occasions, religious celebrations (see for example Oepen, Trudel, Encalada in PART 5). Local newspapers and radio stations could promote the event and support environmental awareness. Video shows on environmental issues could be used as a stimulus to generating ideas. Competitions and prizes — not only for ideas but also for concrete achievements (e.g. largest variety of seeds of a given food crop, most efficient irrigation system, largest area reforested by a community) — would link the event with a general promotion of environmental awareness and capacity. Special events tend to attract a large number of people, especially in isolated areas where gatherings are relatively rare. A special event which incorporates fun, entertainment and competition is likely to receive great visibility. Such an event would provide an opportunity to inform and educate, and to gather and discuss local perspectives and concrete options for action.

Primary Environmental Care

Primary environmental care (PEC) projects combine local environmental care with meeting local needs. The projects would be run by local organized groups and could be assisted by or linked with the environmental initiative in several ways. For instance, some staff of the environmental initiative can act as "matchmakers" to assist local groups in obtaining the inputs which they themselves identify as being crucial for projects to succeed. Such inputs may include credit, specific technologies, political support, training courses, networking with similar projects or study visits, as well as specific information and advice. In some cases a revolving fund can be established to support the best community-generated projects that meet PEC criteria. This is particularly appropriate when capital is available to support both environmental and people's welfare. PEC projects build local confidence and strengthen the capacity and skills of local organizations. When they are closely associated with the environmental initiative, they effectively enhance the local support and thus the sustainability of the initiative itself (see for example Trudel, Padua in PART 5).

Linking Benefits with Efforts in Environmental Initiatives

Mechanisms should be identified that proportionally reward the efforts of individuals or groups in the environmental initiative (see for instance Oepen, Trudel, Smith in PART 5). Efforts include all contributions: labor, land, equipment, expertise, as well as costs borne etc. The mechanisms could be ongoing (such as assurance of tenure, or payments for environmental tasks on the basis of the obtained environmental results), or time-specific (such as a prize or reward for particular achievements). Culturally relevant mechanisms of rewarding merit (e.g. ceremonies and public recognition) may be appropriate. This approach reinforces the message that the contribution of local stakeholders to the initiative is important, noticed and valued. It also creates a built-in positive reinforcement of good practices. There are, however, two potential problems that need to be considered. Firstly, the contributions people make towards the initiative need to be seen in proportion to what they are able to give. For instance, while wealthy people may contribute a great deal of money, this may only be a small sacrifice for them. A mother who offers some of her time after working in the fields and looking after a house and family contributes much more in relative terms. The second potential problem is identifying who contributes what. This is especially difficult in cultures where people work mostly in groups. In such a case, it may be more appropriate to reward a group or an entire community rather than individuals.

Integrating the Initiative with Local Empowerment

Participatory assessment and planning processes can deal with initiatives in natural resource management, local welfare and population dynamics in an integrated fashion. Local authorities may be lobbied to enhance local capabilities for income generation, job training, basic education, reproductive health and family planning, and to facilitate a good measure of local awareness and control of local migration phenomena. Poverty, disease and rapid changes in local population have a powerful effect on the management of resources. If the environmental initiative is not concerned with local welfare, health and population dynamics, it may become incapable of dealing with phenomena such as deteriorating quality of life and inequitable distribution of resources. These are often at the root of the opposition and conflicts that undermine the sustainability of environmental initiatives. This option does not at all imply that the initiative should become directly involved with providing family planning services, health care or income-generating opportunities. It does, however, suggest that the initiative help local stakeholders (including government authorities) to consider and discuss resource management issues together with issues of local welfare and population dynamics. Once the relevant actors (e.g. government agencies or NGOs working with local people) have decided what they wish to do about these issues, the initiative may support them to take appropriate action (see for example Oepen, Trudel in PART 5).

Managing an Environmental Initiative

Problem Identification

Identification relates to two very distinct problems which must be treated in the same context: First of all, one must accurately define the existing environmental problem to be solved, and secondly the communication problem, which is

"If I had one hour to solve a problem I'd use 45 min to think about it, 10 min to investigate potential solutions and 5 min to implement the solution" (Albert Einstein)

equally important and in many cases is one of the reasons for the environmental problem. It is important to be aware of this difference in order to design an appropriate communication strategy. Two examples may illustrate this:

- **Waste problems in industrialized countries** The limited capacity of existing dump sites is very often a central problem for cities and municipalities. Hence, formulated positively, the objective could be a significant reduction of the total amount of waste in order to gain time to develop alternatives or to extend the lifetime of existing dumps. The communication problem is that private households hardly know anything about the composition of domestic waste or the possibilities of preventing and thus reducing it. The objective of a communication strategy could be to provide information on the composition and practical ways of separating waste. This type of communication strategy is quite complex as many different target groups must be addressed.
- **Health problems in developing countries** Water-related diseases account for about 80% of all diseases in developing countries. This is due to the inadequate or non-existent disposal of contaminated surface water used for cooking and washing. The link between water quality and diseases is frequently unknown. Thus, the communication objective could be to provide information on the relationship between water and disease and to show practical ways to avoid this dilemma.

The examples indicate that an accurate definition of the environmental problem and its translation into an objective is of utmost importance, as the intrinsic communication goal can be derived from it (see for instance Oepen, Smith in PART 5). For example, in the garbage dump case above the objective to solve the environmental problem could have been the establishment of new garbage dumps. The communication problem would then have been completely different, such as attempting to achieve greater consensus concerning newly identified landfill sites.

So - what is the best procedure? The waste example presented above will serve to demonstrate the different aspects of problem identification. The first step is to accurately define the true environmental problem to which we wish to draw attention. As perceptions of the problem are likely to be very different, this should be done jointly with consumers, administration, politicians and polluters. The environmental problem – undoubtedly - consists in the greatly varying composition and toxicity of the waste to be disposed of or treated properly. Appropriate landfill sites as well as generally accepted ways of treating the waste are difficult to find or to enforce. Hence, the second step follows: a positive goal must be formulated to deal with the environmental problem. Given the significant potential for waste prevention, the total amount to be disposed of must be reduced. However, a number of obstacles impede environmentally sound behavior: traditional disposal habits, the individual valuation of waste, lack of knowledge and alternatives, time resources etc. A communication strategy addressing these complex patterns of behavior must be elaborated.

And how about the communication goal? A Knowledge-Attitude-Practice (KAP) analysis could be very helpful. Such analyses can reveal that general knowledge is relatively high, the attitude towards the environment is generally positive, but the environmental behavior neither corresponds to the knowledge nor does it reflect the existing positive attitude. The relationship between knowledge and behavior thus shows little correlation. With regard to the existing environmental problem, one can proceed on the assumption that knowledge about the relevance of individual action is quite low. Therefore, one communication goal could be the dissemination of relevant knowledge by the media. This would be a prerequisite for a change in environmental behavior but is in itself insufficient. A strategy concentrating on factors influencing environmental behavior positively would be more promising. Among others, economic incentives such as sanctions, compensation in the form of reduced waste taxes as well as improving the possibilities for individual action such as by provision of waste containers for different types of waste should be mentioned. Support by informal social networks, clear messages and the perception of positive consequences of one's own behavior are of enormous relevance.

Hence, communication goals can be very different, depending on their points of departure. The question arises: What factors or combination of factors have the biggest impact? The integration of existing networks into environmental education and information programs has proved to be a particularly effective method to promote environmentally sound behavior. NGOs concerned with environmental and nature conservation , chambers of industry and commerce, schools and political parties come first when a waste campaign is launched. The communication objective could be the social upgrading of waste separation ("Waste is a resource! Don't waste it!"), and to combine it with positive images. It would be worthwhile to recruit popular "local heroes" from sports and music for such a campaign. The probability of success is quite high for two reasons. Unlike the climate relevance of car emissions, for example, waste is an immediate, visible problem for any individual. Secondly, local communities already represent social networks from which one cannot hide without risking sanctions. This increases the social pressure to conform with others, for example to share the separate garbage containers. The basic key to the success of a campaign is that the messages be clear. This calls for joint action by all actors, e.g. relevant divisions of municipal administration as well as public organizations. Waste separation by households in German cities is a good example of effective campaign achievements aimed at the reduction of waste through recycling.

Integrating Local Culture and Traditions with the Environmental Initiative

Traditional beliefs and values should be linked with the objectives of the initiative, expanding and enhancing positive traditional activities (see for instance Rogers et al, Oepen, Trudel in PART 5). For example, some resources could be dedicated to collecting background information on traditional practices and activities, and these could be discussed in joint meetings between local people and project staff. In agreement with the local people, their stories and myths on environmental issues could be recorded and stored in ways that provide easy access (e.g. cassette tapes). The recordings could be presented to the community as a contribution from the environmental initiative.

Continuous Communication Program

The task of maintaining continuous relationships with local stakeholders should be assigned to experienced staff (see for example Rogers et al, Oepen, Trudel in PART 5). They could assist them in primary environmental care initiatives and other projects to generate benefits and economic returns from environmental protection. For instance, relationships could be maintained by a regular series of events such as a weekly or monthly radio program, or a theater group performing at ceremonies or local social occasions in which people expect to hear news about the environmental initiative. Events should be made as interactive as possible by accepting calls from listeners, reading out letters received, inviting local speakers, asking the audience to comment, intervene in the scene, etc. A regular newsletter in the local language is another possibility if it is comprehensible to local people and addresses matters of interest to them. This can be achieved by involving local people in the preparation of the newsletter and other events to enrich and "test" the effectiveness of the chosen communication tools and avenues. Other systems of communicating information appropriate to the area and the initiative could include pamphlets and posters, presentations to schools and churches, guided tours of the environmental area etc. It is important that the methods used to communicate take into account the needs of those who are illiterate. In this sense, posters, photo stories and audiovisual media are particularly appropriate. Ongoing communication is important for the maintenance of trust between the parties. The links also facilitate a sharing of information and the prevention of conflicts. However, being "in touch" is not enough. As issues arise, the management of the environmental initiative needs to respond to local concerns and take action as appropriate.

Networking with Local Leaders and Opinion-makers
Local leaders and opinion-makers should be kept informed about the environmental initiative through meetings, letters, telephone, personal contact etc. (see for example Oepen, Trudel, Padua, van Woerkum et al in PART 5) This will reduce the possibility that they resent the environmental initiative and use their influence to undermine its credibility. In some areas, established traditional organizations may be more appropriate than new organizations with less credibility and relevance to local communities. In other areas the opposite may be true. There may also be conflicts of power between the old and new, with each trying to establish or retain its power base. Local people should be consulted before assuming which organizations have the most relevance and usefulness to the management and implementation of the initiative. It is crucial to remain outside local power struggles.

Sharing Responsibility through Cooperation

Inter- and Intra-institutional Cooperation
Ultimately, all communication planning is intended to result in action. The major goal of EnvCom is improved environmental quality and increased environmental protection by promoting ecological sustainability. This implies behavioral change. As numerous studies have shown, improved environmental knowledge and changed attitudes are not sufficient to bring about behavioral change. Environmentally sound behavior is influenced by a number of other motives as well as by the institutional framework. Inter- and intra-institutional cooperation is therefore essential, if we focus on behavioral change. Coopera-

tion, not competition, will lead to better results in the long run (see for example Rikhana et al, Oepen, Encalada in PART 5). This is particularly true of environmental problems, which are highly complex and cannot be solved by individuals or a single organization. The implementation of the concept of sustainable development shows this impressively: we need cooperation between different actors on different levels. Therefore, a central precondition for an EnvCom strategy to be effective is cooperation in and between the institutions involved. This would appear to be self-evident but in reality well-meant EnvCom strategies very often fail at this point. However, if cooperation is so important, why does is not come about, where are the barriers and how can these barriers be overcome?

Introducing the catalyst for cars in Germany

The relatively fast and uncomplicated adoption of the catalyst in Germany is a good example of inter-institutional cooperation. Media campaigns by the Ministry of Environment and the automobile associations went together with new bylaws related to pollution, different gas prices and car taxes as fiscal incentives, and cooperative behavior on the part of automobile industries, which at a very early stage anticipated the new technology. A communication strategy, exclusively implemented by the Ministry of Environment would have influenced the knowledge and the attitudes of consumers, but might not have had a great impact on their behavior. Only the cooperation between different institutions with the same vision and target made this success happen.

The reasons for the lack of cooperation are manifold and generally structural ones:

- knowledge is power and sharing information is often perceived as loss of power,
- institutions do not have a mandate to cooperate,
- established ways of cooperation do not exist,
- differences in power - existing or perceived - between institutions make the weaker one appear not to be of equal value,
- potential partners are unknown due to a more introverted behavior of the institution,
- mistrust leads to the "enemy-is-out-there" syndrome, where other institutions are perceived as enemies instead of potential partners,

- an old-fashioned understanding of accountability ("I have to do everything myself!") as well as the fear of losing control ("I have to have everything under control!") impedes cooperation.

The reasons all have a structural deficit in common: The lack of incentives makes cooperation in and between institutions on an individual as well as on an institutional level not appear worthwhile, sometimes even dangerous. This is the case even if cooperation is considered necessary by the people involved. Additionally, the complexity of a problem impedes cooperation. Normally the implementation of projects is bound to partner organizations which are at the same time the target group of the intervention. Strengthening partner organizations means among other things the improvement of internal and external relationships. The strategy should consist in demonstrating the individual and collective advantages of cooperation and helping organizations establish corresponding incentive systems; strengthening the corporate identity thus resulting in a better position in the institutional landscape, improved possibilities for raising new funds and an enhanced reputation.

The improvement of inter-institutional cooperation is somewhat more complicated since intra-institutional cooperation is very often a prerequisite. It is an advantage, in principle, if projects agree upon having a diversified partner structure right from the beginning. This facilitates cooperation with different actors later on. Here it is also important to create incentives. Projects with external advisers can take strategic advantage of their special status. Generally, they are not bound into hierarchies and consequently can act relatively freely, very often having access to all hierarchies in all institutions. The analysis of preferences and restrictions of all actors and institutions to be involved in a communication strategy supplies them with an information lead. This can be used strategically and utilized for cooperation.

To sum it up, a successful communication strategy is a central precondition for inter- and intra-institutional cooperation. Team-building and vision workshops, training and other similar events can be used to achieve this. Instruments and methods mentioned here apply to all levels of intervention, whether they are used within divisions or between divisions and/or institutions. The types of training are numerous. The fields of training include organizational development, team-building, networking, sustainable development, environmental communication, conflict management, systems thinking and many more. The creation of temporary cross-sectoral working groups dealing with a specific task should be taken into consideration as well. Clear role descriptions and tasks, transparency and openness in dealing with persons' expectations and fears are decisive for success. Joint field or project visits or informal round tables help to identify common interests and to reduce mistrust. A rather unobtrusive way to initiate cross-institutional cooperation is to provide advanced training. In such cases selected persons from all relevant institutions should be invited. The subject of the training does not necessarily have to relate to the issue of the planned communication strategy. The topic "Conflict Management" is particularly suitable for promoting better inter-institutional cooperation. Cooperation and communication as well as expressing personal interests, fears and expectations are at the center of simulations and role plays that are part of the training.

Strategic Alliances with Appropriate Partner Institutions

Strategic partner institutions should be identified which offer training or outreach programs for the majority of target beneficiaries (see for example Rikhana et al in PART 5). However, the compatibility of an institution's mission or mandate with environmental education and communication goals, and the added value resulting from such a synergy should be observed. Environmental initiatives should encourage "buy-in" by relevant agencies or institutions, and capitalize on interagency or multisectoral contributions and participation of multidisciplinary teams of specialists or resource persons. Environmental education and communication requires a multidisciplinary, cross-sectoral, and interagency team approach. Policy advisory committees and implementation task groups may support such an approach. Coordination and collaborative mechanisms can be specified by line agencies concerned, for example, with the environment, agriculture, health, education, population, rural development, etc. The latter should enlist active environmental champions and local development stakeholders.

Partnership and Institution Building

Instead of "retailing" EnvCom activities by organizing and conducting training on an ad hoc, sporadic basis, make "franchising" arrangements by collaborating with interested training institutions in developing their staff capabilities in planning, developing and conducting related activities in a sustainable and institutionalized manner. Efforts should focus on integrating relevant environmental education and communication topics into existing curriculum or regular extension and outreach activities. Master trainers and moderators should be trained in the application of EnvCom methods and instruments. The reproduction and utilization of pretested EnvCom materials as a standardized, generic, "must-know" training contents package should be supported.

Mainstreaming the Environment

Strategic Positioning of Environmental
Education and Communication Messages
In message design, the critical interplay among issues concerning sustainable agricultural development, food security, rapid population growth, environment deterioration and other relevant poverty alleviation issues such as health etc. should be analyzed. Relevant critical messages of population education, environmental education and communication and agricultural extension and training should be matched so as to identify priority contents and integrate entry points (see for example Rogers et al, Rikhana et al, Oepen, Encalada in PART 5).

Communicating Environment Issues
with the Rural Population
Relevant messages can be "piggy-backed" through existing communication channels which have large and regular clientele in an institutionalized and sustainable manner (see for instance Rikhana et al, Trudel, Encalada, van Woerkum et al in PART 5). The suggested target beneficiaries can be master trainers of relevant training institutions, trainers of extension and outreach workers, field workers of public extension services and NGOs, selected farmers and community leaders. The potential partner institutions with which alliances should be built are agricultural extension and training agencies, environmental management agencies, rural development training and adult education institutions, rural development oriented NGOs, community-based development centers or multimedia development, production and training centers. Other suggested communication support activities may include strategic multimedia extension campaigns (SEC), social marketing programs, the use of voluntary workers or intermediaries (e.g. school teachers, religious leaders, village chiefs, health workers, etc.) as "barefoot" environmental education and communication champions, the use of popular culture for shaping public opinion, and facilitating policy and technical exchange through conventional and virtual networking.

Increasing Stakeholdership and
Facilitating Sustainability
Using a participation-oriented approach to encourage local ownership of environmental initiatives and EnvCom strategies, training needs assessment and problem identification methods facilitate strategic planning processes for designing relevant and demand-driven environmental education and communication programs (see for example Rikhana et al, Oepen, Trudel, Encalada in PART 5). Participation of key stakeholders at all stages of the planning, imple-

mentation and monitoring process is critical for transferring technical and management skills and program responsibilities. Participatory curriculum development activities through a series of training module writeshops and content materials packaging workshops by a team of local trainers and a pool of multidisciplinary resource persons can produce more relevant content materials for training and outreach. The use of local trainers and teams of resource persons in developing and testing learner-centered, client-focussed, and needs-based EnvCom modules or materials will facilitate the local ownership and proper utilization of such modules, and the sustainability of related activities. This is more successful than employing expert-driven and top-down oriented training packages, and the common, but often unsuccessful and unsustainable practice of expecting local trainers to use imported training materials developed by internationally known and highly qualified experts which may not be relevant to trainees' needs and the specific environmental education and communication learning objectives.

Inducing Quality Assurance and Standards

A general consensus on the EnvCom conceptual framework, operational process and implementation guidelines should be obtained among cooperating institutions. Objectives, measurable outcome indicators, the implementation time frame and resource requirements should be determined and agreed upon (see for example Rikhana et al, Encalada in PART 5). Opportunity for participatory peer review and progress monitoring is essential during critical stages. If available, constructive competition and consultation features through "rewards & recognition" (e.g. participation in international meetings, appointment as regional resource person or consultant, use as model activity, increased funding, etc.) will encourage quality assurance. The same applies to a panel of internationally recognized scholars and practitioners in education and communication to provide independent quality assessment, identify best practices, and give advice for further improvement, expansion or replications.

Facilitating EnvCom Replications

In order to prepare a proper documentation of EnvCom initiatives, the process and critical decision-making steps as well as the implementation procedures and the contextual background should be documented in detail. Standardized performance, outcome and impact indicators for comparative analysis of EnvCom activities facilitate this effort. In order to complete the "last-mile" tasks of consolidating, summarizing, and disseminating the process, methods, results, and lessons learned of EnvCom activities in a user-friendly, attractive and captivating manner, adequate resources have to be committed. These efforts are especially aimed at relevant policy and decision-makers, for further improvement, expansion and replications of EnvCom activities at various levels (see for example Rikhana et al, Oepen in PART 5).

Checklist for Environmental Communication in Projects

Fields in which EnvCom is particularly useful

- support and capacity development of environmental centers, institutions and administrations,
- urban-industrial environmental protection
- development and implementation of national, regional or local environmental action plans or sector strategies
- environmental management in rural regional development planning, social forestry or park management
- solid waste management, energy and water management

Preconditions for Effective EnvCom

- generate technical know-how
- integrate EnvCom up front in project planning
- provide advanced training options
- allocate appropriate staff and funds
- define EnvCom as an output (supporting the goal of a project, e.g. "Information on EIA law disseminated") or an activity (supporting the output of a project, e.g. "Communication strategy on recycling developed with relevant actors")
- plan the communication strategy ahead, taking research, continuous M&E, process documentation and an exit strategy seriously up front
- start locally and on a modest level, and link issues raised, problems addressed and solutions proposed to existing trends, services and potentials
- make use of upstream compatibility of media, e.g. theater - video - TV, if possible by "piggy-backing" one on the other
- diversify the operational levels, e.g. local theater, city newspaper and national TV
- use participatory approaches in media production, management, training etc. to increase local ownership and, hence, program effectivity, significance and sustainability

Potential Fields of Consultancy

- advanced training in specific EnvCom strategies
- training in selected EnvCom methods, instruments and media
- development of a local pool of experts
- process coaching during EnvCom implementation
- exchange of experience at the international, national, regional and local level
- capacity and institutional development
- strategic alliances (partnerships, twinning etc.)

EnvCom Consultant Profile

- needs-oriented and participatory
- more process- than goal-oriented
- incorporation of local know-how and partners
- communication or social science
- participatory methods of EnvCom
- media design
- conflict management and mediation
- interdisciplinary cooperation
- strategic and systemic thinking
- moderation and visualization skills
- process coaching in an intercultural context
- capacity and institutional development

PART 7 - Resources

Web-based Environmental Communication Training - Reflections from Work in Progress

Thomas Zschocke

In his landmark publication "Deschooling Society", Ivan Illich (1972) proposed the establishment of what he called an "educational web" as a radical alternative to formal education. This network would allow learners to engage in creative and exploratory learning with others motivated by the same concerns. Almost thirty years after Illich described his vision, the global network of computer networks the Internet and its graphical interface the World Wide Web seem to provide the technical infrastructure that can be utilized to realize Illich's "educational web."

In general, computers can help us to extend our understanding of the environment and the effect of our behavior on it (Young 1994). As tools for data collection, storing and organizing, computers can be utilized to monitor industrial and natural processes in the global environment and to model simulations of experiments that otherwise would be too dangerous or time-consuming to be conducted in a real setting. And, of course, computer networks can help to connect those who are working on environmental issues in different parts of the world.

Computers have also been widely used in environmental education as an instructional tool. As Rohwedder (1990) points out, computer-aided environmental education has the potential to provide students with a learning experience that demonstrates the complex interaction of environmental systems as well as the impact of our behavior on the environment in a visual and interactive format. Notable examples of computer simulation games for education about environmental issues include "Fish Banks, Ltd." and "Strategem" by Dennis Meadows (see URL below and Martin in this PART) as well as "Ecopolicy" by Frederic Vester (see URL below) or "Moroland" developed by Universität Kassel (see von Papen 2ooo).

There is a growing trend to utilize international computer networks as an infrastructure for distance education, where learners and instructors communicate and cooperate directly with each other and with experts in the field around the world (Filho and Tahir 1998). These computer networks also provide access for people active in environmental communication and education to relevant environmental information. Selected relevant on-line resources include Unesco's International Directory on Environmental Education Institutions, GreenCom, the Environmental Education and Communication Project of the U.S. Agency for International Development, EcoNet of IGC Internet, a member of the Association for Progressive Communications, the Biennial Conference on Communication and Environment - COCE, and the Environmental Communication Resource Center at the School of Communication of the Northern Arizona University (URL see below). The author has put together and annotated the most important information resources about Environmental Communication on the Internet and links to related government agencies, consulting companies, research institutions, NGOs, and other interested parties on a Web page. The page serves networking activities and information exchange in this relatively new field (URL see below).

In fact, all these resources are provided on the Web. Since its inception in 1989, the Web has grown exponentially worldwide. It has now become a driving force for new and innovative approaches in business (e.g. electronic commerce) as well as in education (e.g., Web-based training). The Web allows for the integration of various media types ranging from plain text to streaming audio and video in a single distributed hypermedia environment. Users have the ability to access materials of and contribute materials to an environmental knowledge base at any time and from any place where they have an Internet connection. And, most importantly, the Web also gives users the ability to share their ideas and concerns with others through environmental computer-mediated communication (CMC), primarily computer conferencing (McMahen & Dawson 1995). And, for those who do not have access to the Web or whose access is too expensive, there are services available that forward the contents of Web pages by e-mail.

Although accessing and running a Web site require some investment in human resources and the physical infrastructure, the underlying language of the Web, i.e. hypertext markup language (HTML), and most of the standard Web browsers or clients, i.e. Netscape Navigator and Microsoft Internet Explorer, are available free of charge. Combined they represent a set of low-cost tools for developers of hypermedia applications for environmental communication training as compared to costly multimedia authoring tools such as Macromedia Director or Asymetrix Toolbook. With HTML instructional designers can produce Web pages with any word processor. Such a Web page can include text, tables, forms and graphics. As HTML only allows for structuring Web documents, the visual layout and typography of such a document can be enhanced by using Cascading Style Sheets (CSS), i.e. commands for a set of rules that can be embedded in a Web page using any word processor.

In order to increase the interactivity of such a Web site, an instructional designer can use a programming language like JavaScript. This scripting language allows for the manipulation of properties on the client-side by responding to events or input from the user. Server-side applications (e.g. back-end databases) can be designed for instance with the programming language Perl that produces Common Gateface Interface (CGI) scripts. And, stand-alone applications of smaller size called applets can be developed with the programming language Java. All of these programming tools are also available free of cost.

Finally, an instructional designer can take advantage of helper applications in the development of Web-based environmental communication training. When embedded in a Web page certain media applications cannot be represented by HTML and launch instead a helper application that resides on the user's computer. For instance, a trainer could include a Word or a spreadsheet document in the Web page. The browser would automatically launch the appropriate program (e.g., Microsoft Word or Lotus 123). If the user does not own these applications, now additional viewers are available that can be installed on the user's machine at no cost (e.g. Microsoft PowerPoint viewer for Power-Point slide presentations). In other cases the designer might have included a multimedia application, such as streaming video or a shockwave movie (i.e. Macromedia Director or Flash file), then the browser would automatically open a plug-in that displays this particular application (e.g. Quick Time or Shock-wave/Flash player). Although developing these kinds of applications requires that the instructional designer own the developing tools, most of the helper applications at the user's end are available free of charge..

To summarize, the Web allows for the delivery of a great variety of multimedia applications in a single environment. In utilizing these possibilities for training, the Web represents an additional method for delivering instruction along with video, CD-ROMs and workbooks. The Web can be best used as a performance technology to help learners gain more knowledge or cognitive skills such as problem-solving, applying rules or distinguishing among items. This can be best achieved by using text, graphics and symbols as well as instructional strategies such as reading, writing, completing exercises or solving computational problems (Driscoll 1998).

As in any training event the content and interaction during the training sessions are centered on the learner's needs. In designing a Web site the same learner-centered approach needs to be applied so that learners can identify with the content and draw on their experiences. The lack of face-to-face meetings can be replaced by computer-mediated communication, e.g. e-mail, threaded discussion lists or chat rooms. Although, as Driscoll (1998) points out, Web-based training requires the necessary technical infrastructure and computer skills on the part of the learner, it allows trainers to provide training when and how it is needed, it reduces the costs of reproducing, packaging, mailing as well as updating training material, and it provides access to subject-matter experts online as well as to knowledge databases and other selected audiovisual material.

Currently the author is developing a module for environmental communication as part of his research on Web-based training at the Center for International Education, University of Massachusetts Amherst, U.S.A. This training module is designed as a supplement to existing material for the training of middle-management practitioners in environmental communication carried out by Manfred Oepen at ACT - Appropriate Communication in Development in Germany. In close collaboration with ACT, the author is designing a training module that combines resource information about and Web links to environmental communication with interactive applications that simulate the process of an environmental communication campaign.

In an introductory section the learner will be exposed to the basic concepts of environmental communication, its media and approaches of message design (e.g. Parker 1997) and additional resources. An important element of this training module is an interactive application that is based on the "10 Steps Towards an Effective Communication Strategy" outlined in PART 4. A brief version of this 10 step-strategy is also available as a practical orientation brochure by GTZ and OECD (GTZ 1999, OECD 2000), both also available as pdf-files that can be downloaded from the respective web sites.

Different scenarios and case studies will help to illustrate the process and help the learners to apply their experience to the training exercises. In addition, pictures and possibly video clips will provide additional means of illustration. The core element of the training module is an interactive tool which simulates the decision-making process when preparing an environmental communication campaign. This simulation is based on a decision and planning model developed by Adhikarya and Posamentier (1987). In this model a given situation is described based on the different skills and knowledge of the participants in the environmental communication process. Based on their characteristics, different strategies may evolve leading to a specific set of goals and the utilization of certain kinds of media to support the environmental communication campaign. In the Web-based training module learners are confronted with an

"If – Then" situation where they have to analyze a given situation as the foundation for their proposed environmental communication campaign. Finally, the Web-based training module also supports e-mail so that subject-matter specialists can be consulted during the use of the module. Links to additional resources on the Web help to connect the learner to more information and other organizations and experts that can be contacted to address specific questions or problems of the learners. Of course, this Web-based training module can also be used off-line with any standard browser that supports HTML, JavaScript and Cascading Style Sheets (i.e. Dynamic HTML or DHTML). However, the communication tools will not be available.

In addition to the Web-based training module, the author is developing a database with a collection of initially more than 30 games and exercises for environmental management in collaboration with ACT - Appropriate Communication in Development and the German Foundation for International Development (DSE). The database is designed as a resource for facilitators of training activities in environmental management. A demo version is currently being tested (URL see below). Delivered as an interactive multimedia application on CD-ROM, the games in the database can be searched for in alphabetical order or by category. The individual games contain a short description, learning goals and objectives, materials, target audience, duration, procedure, and additional comments and remarks. The database will help facilitators to quickly review and select appropriate games and exercises for a specific training situation, and read detailed instructions on how to implement them best.

It is hoped that the Web-based training module for environmental communication along with the database of games for environmental management will provide a useful supplement to ongoing training activities in this field and contribute to the development of an "educational web", as Illich proposed.

Internet addresses of information mentioned in this article

"Fish Banks, Ltd." and **"Strategem"**
> by Dennis Meadows:
> *http://www.unh.edu/ipssr/Lab/FishBank.html*
> respectively /Stratagem.html

"Ecopolicy"
> by Frederic Vester:
> *http://www.frederic-vester.de/ecopolic.htm*

Unesco's International Directory
on Environmental Education Institutions
> *http://www.unesco.org/education/educprog/environment/*

GreenCom
> the Environmental Education and Communication Project of the U.S. Agency for
> International Development:
> *http://www.info.usaid.gov/environment/greencom/*

EcoNet of IGC Internet
> a member of the Association for Progressive Communications:
> *http://www.igc.org/igc/gateway/enindex.html*

Biennial Conference on Communication and Environment - COCE
> *http://www.yorku.ca/faculty/academic/meisner/coce/index.html*

Environmental Communication Resource Center
> School of Communication of the Northern Arizona University:
> *http://www.nau.edu/soc/ecrc/*

Annotated Bibliography and Links on EnvCom on the Internet
> by Thomas Zschocke:
> *http://www.umass.edu/wbt/envcom/*

Web-based EnvCom Training Module
> by Thomas Zschocke:
> *http://www.umass.edu/wbt/*

Environmental Communication for
Sustainable Development. A Practical Orientation
> by GTZ:
> *http://www.gtz.de/pvi*

Environmental Communication
> Applying Communication Tools towards Sustainable
> Development by OECD;
> *http://www.oecd.org*

Data Bank on Games and Exercises related
to Sustainable Development
> for ACT and DSE by Thomas Zschocke:
> *http:// www.umass.edu/wbt/dsela21/*

Linking Key Concepts: Team-building, Systems Thinking and Sustainable Development
Gillian Martin Mehers

Trends in sustainable development training are widening from those which focus solely on technical environmental aspects, to include more process-related topics such as environmental communication, environmental impact assessment, participatory mechanisms, environmental conflict resolution and consensus-building and environmental decision-making. Training on these process tools focusses more on the holistic view of sustainable development than on purely sectoral needs. This trend is beginning to reflect the systems' principle of interrelatedness. A large body of research already exists in the adult education field about how adults learn and the most effective educational techniques required to reach them. The application of this knowledge to environmental management training is not, however, as common practice as would be expected. The environmental education community is growing. Science education departments in universities are expanding, and institutions of higher education are developing curricula for courses on environment, as well as "greening" their curriculum to integrate environment and development issues into traditional fields of study. The training work with mid-career environmental professionals, however, has not always received the same pedagogical attention as the formal education system. Environmental protection, that is the prevention and mitigation of environmental problems, in the past has been seen as a technical/scientific field. Trainers, therefore, have not always been educators, but technical experts transferring technical knowledge. "Fixing" the environment, however, is not like fixing a car. The capacity needed to understand and manage the environment includes developing the ability to incorporate the social dimension, and to ensure the personal involvement of important stakeholders in the joint task of amelioration and management of the environment, and to ensure a multidisciplinary approach.

Capacity-building for environmental management and sustainable development presents a particular challenge for those working in the field. Capacity-building for the implementation of Agenda 21 refers to the process by which the human and financial resources of a country's people and institutions are developed, enhanced and utilized to identify and follow the country's own path towards sustainable development. The capacity-building concept goes beyond the scope of institutional strengthening, which has been the spearhead of development assistance over the past 25 years. It is not limited to skill development through isolated training activities. Instead it resembles more the processes followed by learning organizations, encompassing the identification of inefficiencies in the way institutions operate to achieve defined sustainable development objectives, and the transformation of people and institutions to implement changes which are needed to improve performance. To this end, it is proposed here that three key concepts should be linked in the context of sustainable development training and capacity development: Team-building, systems thinking and sustainable development.

Team-building   Performing is one of the final stages of team development. One of the objectives of any training or capacity development process is to help teams reach this stage. Prior to this, the task of the highly effective team is to define its goal, share this vision, and work together to achieve it. The process of defining and sharing the goal is as important as the performance itself. It is through this phase of activity that the team can build and practice skills in, for example, communication, decision-making, problem-solving and conflict resolution. It can explore diversity and understand different personality profiles for both leadership and team members. Related training activities should lead the team through the phases of group development, introduce and practice these skills and move a team into the performance stage in the process. Tools such as personality profile questionnaires, brainstorming and active listening activities, problem-solving initiatives, and exercises on giving and receiving feedback could be used.

Systems Thinking   Systems thinking is one tool a team can use to understand patterns of interaction, for example, among team members, the functioning of their organization, and its relations within a larger context. Equally importantly, it forms the bridge between sustainable development and patterns of interaction at a more global level. Games help show systems dynamics in an experiential way, before lectures and discussions show the application of the activity through analogy. In a training event, team members can also experience a system through a game which is played called "Fishbanks." (Meadows 1991). This is a computer-assisted game where teams, representing fishing companies, try to manage the renewable resource of fisheries in the coastal and deep sea off an imaginary coastline. Caught in the competition of the game, teams can see how easy it is to deplete fish stocks irrevocably, when knowledge about the system is missing, collaboration between fishing companies is lacking, and strategies for together sustainably managing the resource do not exist. Systems thinking vocabulary becomes common use in the training as team members begin to share their observations of patterns, levels of perspective and can apply loops such as addiction and adaptation to everyday forms of behavior in

264

their organizations or personal lives. Among others, "interrelatedness", "reference mode" and "time frame", and "eroding goals" become concepts which can be used in conversation, which Peter Senge says is the "single greatest learning tool in (an) organization – more important than computers or sophisticated research." (Senge 1994:14)

Sustainable Development In addition to the lessons for management of sustainable development which emerge during the systems thinking activities – such as the "tragedy of the commons" and "growth against the limits" – sustainable development training and capacity development should provide time for the team to create its own future goals in relation to work towards sustainable development. Using the team-building and systems thinking components, action planning sessions should allow the team to practice and plan their performance, set realistic goals and plan the next steps to reach them.

Appendix

Selected Environmental Communication Training Opportunities

Environmental Communication for Sustainable Development
by ACT - Appropriate Communication in Development

Sustainable development depends on the active participation of the intended beneficiaries because development requires people to consciously and willingly change their behavior. Environmental communication (EnvCom) is the planned and systematic use of simple audiovisual and sophisticated electronic media among planners and beneficiaries of environmental initiatives. The actors concerned improve their social and communicative competence for an enhanced impact of local self-help and environmental management services. EnvCom helps to design projects and action plans successfully through "dialogue" between planners *and* the people concerned, and mobilizes them for environmental action. International experience indicates that the probability of success for environmental initiatives can be increased if the elements "EnvCom", "cooperation" between government services, NGOs and media, as well as "mediation" are combined.

The general learning objectives for this curriculum on "Environmental Communication" have been met if training participants

- are aware of participatory forms of communication and media that should be integrated in the different phases of a program or project cycle of environmental initiatives,
- have realized the potentials and limits of cooperation and networking with NGOs and other parties involved in the process of environmental initiatives,
- have gained an overview of basic elements and methods in public relations and media use as well as the differences in use of various media, forms of communication and other instruments regarding their relevance and usefulness for environmental initiatives,
- can point out potential contributions to constructively overcome conflicts or sociocultural disparities by means of appropriate media, public relations and cooperation with NGOs and other parties,
- are able to identify specific action-oriented applications for integrating NGOs and participatory media in long-term environmental initiatives within their own working space.

The methodology is based on three major working principles - moderation, visualization and experiential learning. The principles are applied throughout each workshop, building on participants' know-how and experience. The major features are:

- transparency in decision-making,
- needs-orientation,
- cooperation and participation, i.e. each idea counts so that there is fair dialogue without domination or hierarchies,
- learning by self-reflection, encouraging participants to reflect on their own behavior in group work, exercises, role plays etc.
- 10/60/30 rule, i.e. according to experience in adult education theoretical inputs are limited to a minimum (10%). A lot of time should be allocated to fieldwork, exercises, group work, role plays etc. (60%). Sufficient time should be reserved for discussions and conclusions (30%),
- professional moderation and visualization,
- evaluation ensures a shared responsibility among all participants.

Six modules (**1** - Introduction, **2** - Basic Concepts, **3** - Communication Strategy, **4** - Case Studies, **5** - Practice-oriented Life and **6** - Case and Evaluation) and up to 23 learning units constitute the tailor-made course contents.

Target Group	Middle management and field staff of NGOs and other agencies
Duration	Modules of 1, 3, 14 and 3o days' duration are available which can be modified on request (e.g. over two weekends)
Language	English or German
Next Course	on request
Place	on request
Costs	on request
Contact	ACT • c/o Manfred Oepen • Kleine Twiete 1 A 30900 Wedemark, Germany Tel +49 - 5130 - 60 130• Fax 798 03 Email act.oepen@t-online.de

Community Communication for Rural Development

By the German Foundation for International
Development - Center for Food and Agriculture,
DSE - ZEL, Feldafing

The training methodology is based on the same working principles as the previous one - moderation, visualization and experiential learning. The basics of development, communication and participation concepts in **Module 1** help to bring out the advantages of self-experience, cooperation and participation. In **Module 2,** the basics of community communication are elaborated through participatory methods in the 1o steps towards an effective communication strategy. In **Module 3,** the participants evaluate the effects of various media. At the heart of the workshop, newly gained skills are applied to practice in rural "real life" situations in **Module 4.** "Transect walks", "focus-group discussions", "mapping" and other instruments are used in cooperation with local groups. Participants design a communication strategy, select, pretest and produce media. A critical evaluation of the communication strategy and the workshop contents and methods close the course in **Module 5.** The course was developed by communication specialists and was held with participants from seven African countries in Ghana in September 1994 and from nine Asian countries in Indonesia in May 1996.

Target Group	Middle management of government and nongovernmental organizations
Duration	One month, 7-10 days on request
Language	English, German or other languages on request
Next Course	on request
Place	on request
Costs	on request
Contact	DSE - ZEL • Wielingerstr. 52
	82 336 Feldafing
	Tel +49 - 8157 - 938 100 • Fax 938 777
	Email pfeiffer@dse.de

Environmental Communication

by the German Foundation for International
Development - Center for Public Administration,
DSE - ZÖV, Berlin

In **Module 1**, the training comprises introduction rounds related to participants, workshop goals and working principles, followed by the basics of sustainable development, communication and participation concepts in **Module
2**. **Module 3** elaborates the didactics of environmental communication by means
of a 10-steps communication strategy. Different media are compared and their
"participation potential" evaluated. The analysis of potentials and performances of particular communication processes and media are intensified through
case studies in **Module 4**. At the heart of the workshop, in **Module 5,** the
newly gained skills are applied to a "real" field study in cooperation with social
groups facing an environmental problem. A critical evaluation of the communication strategy and the training contents and methods close the course in
Module 6.

The training manual comprises

- six practice-oriented training modules,
- handouts, teaching aids, literature,
- case studies related to EnvCom media and techniques,
- OHP transparencies,
- games, role plays, simulations and exercises etc.,
- background material.

Target Group	Middle management of government and non-governmental organizations, field staff on request
Duration	Two weeks, shorter on request
Language	English, German or other languages on request
Next Course	on request
Place	Vietnam, Indonesia, Zimbabwe and Southern Africa, other locations on request
Costs	on request
Contact	DSE-ZÖV • c/o Hinrich Mercker Heussallee 2-10 • 53113 Bonn Tel. +49 - 2434-224 • Fax 2434-233 Email mercker@dse.de

Environmental Communication
by the Deutsche Gesellschaft für Technische Zusammenarbeit (GTZ) GmbH - Pilotvorhaben Institutionenentwicklung im Umweltbereich, **GTZ - PVI**

Environmental sustainability has become one of the strategic goals in "Shaping the 21st Century". Therefore, many projects are about changing people's impact on the environment. Implementing agencies realize that ownership, participation and empowerment, involve multifaceted communication processes in inter-institutional cooperation, interaction and consensus-building between a wide range of actors.

Integrating communication, social marketing, public relations and non-formal education inputs in project life cycles is at the heart of the matter of a training workshop offered by GTZ. Field studies and the practice-oriented, step-by-step development of a communication strategy as an integral part of project planning will illustrate to participants how environmental communication can become a powerful tool in project management.

Target Group	Middle management of projects and programs
Duration	3-4 days
Language	German, English and other languages on request
Next Course	on request
Place	on request
Costs	on request
Contact	GTZ-PVI • c/o Winfried Hamacher
	Wachsbleiche 1 • 53 111 Bonn
	Tel +49-228-98 53 30 • Fax 98 57 018
	Email gtzpvi@aol.com

Job Aids

The following forms are related to the 10 Steps of an EnvCom Strategy (partly adapted from Martin Mehers 1998). Each form is meant to be used per factor, strategic group or media involved. Therefore, the forms are 'open-ended', and should be adjusted to the user's specific needs.

Job Aid 1   Step 1 - Situation Analysis and Problem Identification

Location : _____

Key Factors	Problem Priority	Alternatives to Problematic Behavior

Job Aid 2   Step 2 - Identification of Strategic Groups

Group	Interest	Critical Behavior

Job Aid 3   Step 3 - Communication Objectives

Strategic Group: _____

Project Objective	
Communication Objectives **Practice** : Which practices do the strategic groups have to maintain in order to meet the project objective?	
Attitude : What attitudes do the strategic groups have to be convinced of so that their practices will change in the above-mentioned direction?	
Knowledge : What knowledge (or information) do the strategic groups need so that they can develop the attitudes mentioned above?	

Job Aid 4   Step 4 - Developing a Communication Strategy

IF:				THEN:					
SITUATION + ACTOR	Position of actors in relation with			Environmental Communication Strategy Priorities			Communication Channel Emphasis for Multimedia Mix		
	Knowledge **K**	Attitude **A**	Practice **P**	Main Approach	Main Purpose	Information Positioning	Mass Media (TV, radio)	Group Media (NGOs)	Interpers. Communication (consult.)
1.									
2.									
3.									

Job Aid 5   Step 5 & 6 - Media Selection and Strategic Group Participation

Media	Advantages	Disadvantages	Strategic Group Participation

Job Aid 6   Step 6 & 7 - Message Design and Media Selection

Strategic Group	Message	with what kind of appeal/approach	Media

Job Aid 7   Step 7 - Message Design per Strategic Group

Strategic Group	Message with what kind of appeal/approach	Message with what kind of appeal/approach	Message with what kind of appeal/approach

Job Aid 8   Step 8 - Media Production and Pretesting

(on a scale from 0 – 10, with 0 = low, 10 = high)

Media	Relevance	Text Comprehension	Visual Comprehension	Motivational Potential	Credibility

Job Aid 9   Step 9 - Media Performance and Implementation

Strategic Group	Expected Change in Practice	Time (Period)	Responsibility	Result	Resources

Job Aid 10   Step 10 - Evaluation

Communication Objective	How do you know you are successful?
Group _____	
Group _____	

276

Selected Literature

Abraham et al. 1990, **Can media educate about the environment ?** in: Media Development, 2/1990, S. 6f

Abril/Olivera 1987, **How Groups Can Make Themselves Alive.** A Group Communications Manual, Manila

Adhikarya et al. 1987, **Motivating farmers for action.** How strategic multi-media campaigns can help, Eschborn: GTZ

Adhikarya R 1996, **Participatory Environment Education through Agricultural Training**: Best Practices and Lessons Learned from Six Asian Countries, paper to a GTZ International Conference on 'Communicating the Environment', Bonn, Germany: Dec 15-19

AIJ 1983, **Community Media,** in: A Course Guide in Planning the Use of Communication Technology, Manila

Albrecht et al. 1987, **Handbuch Landwirtschaftliche Beratung,** Bd 1+2, Eschborn: GTZ

Atkin/Meischke 1989, **Family Planning Communication Campaigns in Developing Countries,** in: Rice/Atkin (eds): Public Communication Campaigns, London, pp 227-232

Balit 1988, **Rethinking development support communication,** in: dcr 3/1988, S. 7f

Baskaran 1990, **The rise of the environmental movement in India,** in: Media Development, 2/1990, S. 13-16

Bauer 1992, **Schwachstellen und Lösungsansätze in Beratungsprojekten** der Technischen Zusammenarbeit im ländlichen Raum, Eschborn/GTZ

Bauer 1994, **Beraten Berater?,** in: E+Z 2/ 1994, 47f

Berrigan, 1979, **The Role of Community Communications,** Paris

Beyer 1988, **Umweltbezogene Aus- und Fortbildung** für Länder der Dritten Welt, Berlin

Bhasin U, Singhal A 1998, **Participatory approaches to message design: 'Jeevan Saurabh',** a pioneering radio serial in India for adolescents, in: Media Asia, 25/1, pp 12-18

Boafo 1989, **Communication and culture**: African perspective, Nairobi

Boafo K 1993 (ed), **Media and Environment in Africa,** Nairobi: ACCE

Bordenave 1977, **Communication and rural development,** Paris

Borrini 1991 (ed), **Lessons Learned in Community-based Environmental Management,** Rome

Borrini G 1993, **Enhancing People's Participation in the Tropical Forests Action Program,** FAO, Rome.

Borrini-Feyerabend G 1997-a, **Participation in conservation: why, what, when, how?,** in: IUCN, Beyond Fences, vol 2, Gland, pp 26-31

Borrini-Feyerabend G 1997-b, **Primary environmental care,** in: IUCN, Beyond Fences, vol 2, Gland, pp 74-78

Borrini-Feyerabend G, Brown M 1997, **Social actors and stakeholders,** in: IUCN, Beyond Fences, vol 2, Gland, pp 3-7

Brown LR et al 1993, **Zur Lage der Welt** - 1993. Daten für das Überleben unseres Planeten, Frankfurt

Bruckmeier et al.,1992, **Trägerentwicklung im Umweltbereich,** Berlin: WZB

Byers BA 1996, **Understanding and Influencing Behaviors in Conservation and Natural Reources Management,** African Biodiversity Series, No. 4, Washington D.C.: USAID

CAP 1983, **State of the Malaysian Environment** - Towards Greater Environmental Awareness, CAP

CAP 1984, **Environment, Development, Natural Resources Crisis** in Asia and the Pacific, CAP

CAP 1989, **Development and the Environmental Crisis.** The Malaysian Case, CAP

CDR Associates 1995, **Colorado, Reaching Agreement**: Conflict Resolution Training for the IUCN, IUCN internal report, Gland (Switzerland).

Cernea MM 1985, **Putting People First,** World Bank, Oxford University Press, New York.

277

Chambers R 1983, **Rural Development: Putting the Last First**, Longman, London.

Chambers R 1992, **Participatory Rapid Appraisal** - PRA, IDS Discussion Paper 311, Brighton

Club of Rome 1977, **Reshaping the International Order (RIO-Report)**, London

Cohen J, Uphoff N 1980, **Participation's place in rural development**: seeking clarity through specificity, World Development 8: 213- 235.

Colle 1993, **The pragmatics of development communication**, paper to a Berlin conference by FU Berlin, ACT and WIF on 'Media Support and Development Communication in a World of Change', Nov 19-20

Colletta/Kidd 1981 (ed), **Tradition for Development**, Bonn: DSE

Conference 1977, **Intergovernmental Conference on Environmental Education**. Report, Tbilisi/USSR, Oct 14-26

Davis-Case D 1990, **The Community's Toolbox**: The Ideas, Methods and Tools for Participatory Assessment, Monitoring and Evaluation in Community Forestry, Food and Agricultural Organization (FAO), Rome

Denkmodell (undated) , **SINFONIE** - Systemic Interpretation of the Nature of Factors Influencing Organizations and Networks in Their Environment, Berlin

Di Giulio RT, Monosson E 1996 (eds), **Interconnections between human and ecosystem health**, London

Dietz et al. 1992, **Communication for Conservation**, in: Development Communication Report, 1/1992 S. 4-6

Doran 1994, **Pulling the Strings**. Spreading the environmental message in Nigeria, in: Outreach, 1/1994, S. 63-67

Dovers SR, Handmer JW 1993 , **Contradictions in sustainability**, in: Environmental Conservation 20 (3), pp 217-222

Drijver C 1990, **People's participation in environmental projects in developing countries**, IIED, Dryland Networks Program, Paper No. 17.

Driscoll M 1998, **Web-Based Training**. Using Technology to Design Adult Learning Experiences. San Francisco, CA: Jossey-Bass/Pfeiffer

DSE-ZEL 1996, **Community Communication for Rural Development**. A Training Curriculum and Manual, Feldafing

DSE-ZÖV 1998, **Environmental Communication**. A Two-Week Training Course Module, Berlin

Dubey et al. 1990, **Indian media fail to take environmental issues seriously**, in: Media Development, 2/1990, S. 23-24

Durning A B 1989, **Action at the Grassroots**: Fighting Poverty and Environmental Decline, Worldwatch Paper 88, Worldwatch Institute, Washington D.C..

Edwards D 1998, **Can we learn the truth about the environment from the media?**, in: The Ecologist 28/1, pp 18-22

Epskamp 1989, **Theatre in search for social change**, The Hague

Epskamp 1991, **Popular theatre and the media**: the empowerment of culture in development communication, The Hague

FAO (undated), **Guidelines on Communication for Rural Development**. A brief for development planners and project formulators, Rome

FAO (undated), **Guidelines on Communication for Development**, FAO guidelines 1, Rome

FAO 1994, **Strategic Extension Campaign**, Rome

FAO 1995, **Understanding farmers' communication networks**. An experience in the Philippines, Rome

FAO 1996 , **Communication for rural development in Mexico**, Rome

Festa 1993, **Alternative video and democratization in Brazil**, in: Lewis (ed): Alternatice Media: Linking Global and Local, Paris: UNESCO, S. 107-116

Filho 1993 (ed), **Priorities for Environmental Education** in the Commonwealth, Bradford, 9/1993

Filho L, Tahir F 1990 (eds), **Distance Education and Environmental Education**. Frankfurt/M.: Lang

Floquet 1992, **Le diagnostic concerte des modes de question des ressources naturelle**. Premiere etape d'une strategie de sensibilisation des populations rurales et des intervenants en matiere environnementale, Hohenheim

Fortner R, Smith-Sebasto N, Mullins G 1994, **Handbook for Environmental Communication in Development**, Columbus/Ohio

Fraser C 1994, **How Decision-Makers See 'Communication for Development'**, Special Report in: Journal of Development Communication, pp. 56-67

Fraser C, Restrepo-Estrada S 1998, **Communicating for Development**. Human Change for Survival, London/New York

Fuglesang D 1982, **About Understanding**: Ideas and Observations on Cross-cultural Communication, Dag Hammarskjöld Foundation, Uppsala (Sweden).

Gabathuler 1991, **Contribution a la methodologie et a la didactique de vulgarisation**, Bern

Gajanayake S, Gajanayake J 1993, **Community Empowerment**: A Participatory Training Manual on Community Project Development, PACT (Private Agencies Collaborating Together), New York.

GFA 1994, **SWOT Analysis and Strategic Planning**, Hamburg

Ghai D, Vivian JM 1992 (eds), **Grassroots Environmental Action**: People's Participation in Sustainable Development, Routledge, London.

Ghai D, Vivian JM 1992 (eds), **Grassroots Environmental Action**: People's Participation in Sustainable Development, Routledge, London and New York.

Ghimire K, Pimbert M 1996 (eds), **Social Change and Conservation**: Environmental Politics and Impacts of National Parks and Protected Areas, UNRISD, Geneva (in press).

Glaeser B 1992, **Natur in der Krise?** Ein kulturelles Mißverständnis, in: Gaia 1 (4), pp195-203

GreenCom-USAID 1997, **What Works**: A Donor's Guide to Environmental Education and Communication Projects, Washington D.C.

GTZ (undated), **Manual for Urban Environmental Management**, Eschborn: GTZ

GTZ 1992, **Der Funke ist übergesprungen**. Fallstudie eines Projekts zur Verbesserung der dörflichen Landnutzung in Burkina Faso, Eschborn: GTZ

GTZ 1992, **Entwicklungschancen sichern**, Eschborn

GTZ 1992, **Handlungsfelder der Technischen Zusammenarbeit im Naturschutz**, Eschborn: GTZ

GTZ 1992, **Solid Waste Management with People's Participation**, Kathmandu/Nepal: GTZ

GTZ 1993, **Nyabisindu - Eine ökologische Landwirtschaftsalternative**, Eschborn: GTZ

GTZ 1999, **Environmental Communication for Sustainable Development**. A Practical Orientation, Eschborn

GTZ-PVI 1994, **Information, Bildung und Kommunikation im Umweltbereich**, Bonn

GTZ-PVI 1996, **Umweltprojekte durch Kommunikation verbessern**, Bonn

GTZ-PVI 1997, **Communicating the Environment**. International Conference Documentation, Bonn

Guerrero SH et al. 1993, **Public Participation in Environmental Impact Assessment**. A Manual on Communication, Manila

Haan G de 1995 (ed), **Umweltbewußtsein und Massenmedien**, Berlin

Haan G de 1997, **Paradigmenwechsel**, in: Politische Ökologie 51, p. 22ff

Haan G de, Kuckartz U 1996, **Umweltbewußtsein**. Denken und Handeln in Umweltkrisen, Opladen

Hauff V 1987 (ed), **Unsere gemeinsame Zukunft**. Der Brundtland-Bericht der Weltkommission für Umwelt und Entwicklung. Eggenkamp, Greven

Heidorn 1993, **Umweltbildung in Sarawak**: Lernziel 'Tropenwaldschutz', in: WWF-Journal 1/93, S.15

Hemert M v, Wiertsema W, Yperen M v 1995, **Reviving links**. NGO experiences in environmental education and people's participation in environmental policies, Amsterdam

Hines JM, Hungerford HR, Tomera AN 1987, **Analysis and Synthesis of Research on Responsible Environmental Behavior**: A Meta-Analysis, in: Journal of Environmental Education, vol. 18, No 2, S. 1-9

Hoffmann 1991, **Bildgestützte Kommunikation in Schwarz-Afrika**, Hohenheim

Hoffmann 1992 (ed), **Beratung als Lebenshilfe**. Humane Konzepte für eine ländliche Entwicklung, Hohenheim

Hollenbach 1993, **Modern Media. Not necessarily appropriate**, in: gate 2/93, S. 7-11

Hungerford H, Volk T 1990, **Changing learner behavior through Environmental Education**, in: Journal of Environmental Education. 21/3, pp 8-21

IIED (undated), **Environmental Education Strategy for Ghana**, Environmental Protection Council, Accra/Ghana (with IIED)

IIED 1995, **Participatory Learning and Action**. A Trainer's Guide, London

Illich I 1972, **Deschooling Society**. New York, NY: Harper & Row

Iozzi L 1989, **What Research?** Part one: environmental education and the affective domain, in: Journal of Environmental Education, 20/3, pp 3-9

IUCN 1991, **Walia**. The Approach. Practical Guidelines, Gland

IUCN 1991, **Walia - The Approach and Practical Guidelines**. Schools Environmental Education Project, Mali of the IUCN Sahel Programme, Gland: IUCN

IUCN 1993, **Education for Sustainability** - a practical guide to preparing national strategies, Gland: IUCN

IUCN 1997, **Beyond Fences**. Seeking Social Sustainability in Conservation, Gland

IUCN/UNEP/WWF 1991, **Caring for the earth**: A strategy for sustainable living, Gland

Jacobson S 1991, **Evaluation model for developing, implementing, and assessing conservation education programs**: examples from Belize and Costa Rica, in: Environmental Management 15/2, pp 143-150

Jacobson S, Padua S 1995, **Systems Model for Conservation Education in Parks**: Examples from Malaysia and Brazil, in: Conserving Wildlife, Jacobson S ed., New York City, pp 3-15

Jax K et al 1996, **Skalierung und Prognoseunsicherheit bei ökologischen Systemen**, in: Verh Ges Ökol 26, pp 527-535

Jaycox 1993, **Capacity building**: The missing link in African development, Reston 1993

Jüdes U 1996, **Das Paradigma "Sustainable Development"**. Nachhaltige Entwicklung im Hinblick auf ökologische, kulturelle, soziale und ökonomische Dimensionen. unpubl. research for Bundesministeriums für Bildung, Wissenschaft, Forschung und Technologie (BMFT)

Kabou A 1993, **Neither poor nor powerless**. Against black elites and white aid, Basel

Keiper 1985, **Internationale Vernetzung**. Die 'Consumer Association of Penang', in: EPK 3/85, S. 21-23

Kidd 1982, **The Popular Performing Arts**, Non-formal Education and Social Change in the Third World, Den Haag

Kommunikation eines Leitbildes 2000, in: Politische Ökologie 17, 63/64

Krimsky et al. 1988, **Environmental Hazards**: Communicating Risks as a Social Process, Dover

Krohn W, Küppers G 1992 (eds), **Ebergenz: die Entstehung von Ordnung**, Organisation und Bedeutung, Frankfurt/M.

Labrador 1988, **Community Communications for Self-help Mobilization** in the Philippines, in: Oepen (ed): Development Support Communication in Indonesia, Jakarta, 38-45

Lass W, Reusswig F 2000, **Worte statt Taten**. Nachhaltige Entwicklung als Kommunikationsproblem, in: Politische Ökologie 17(2000)63/64, pp 11-14

Lewis C 1996, **Resolving and Managing Conflicts in Protected Areas**: Guidelines for Protected Area Managers, IUCN, Gland (Switzerland)

Loebenstein et al. 1993, **Kompensation und Interessenausgleich in der Pufferzonenentwicklung** Bd.I/II, Bonn, GTZ

Lohmeier 1994, **Früher Fachspezialisten - heute Berater?**, in: E+Z, 2/1994, 44ff

Marien M 1996 (ed), Environmental issues and sustainable futures, Bethesda

Martin Mehers G 1998 (ed), **Environmental Communication Planning Handbook** for the Mediterranean Region, IAE-Genf

McMahen C, Dawson AJ 1995, **The Design and Implementation of Environmental Computer-Mediated Communication (CMC) Projects**, in: Journal of Research on Computing in Education 27 (3), 318-335

Merkle 1991, **Umwelt und Gesundheit**. Gesundheitsökologie in Entwicklungsländern, GTZ

Merz G 1993, **Umwelterziehung in Ländern der Dritten Welt** - ein Schwerpunkt der internationalen WWF-Arbeit, in: WWF-Journal 1/93, S. 13f

Meyer AJ 1992, **Environmental Education and Communication**: Pulling it All Together, in: Development Communication Report 76(1992)1, S. 1-3

Mody B 1991, **Designing Messages for Development Communication**, Los Angeles

Mody B 1991, **Designing messages for development communication**: an audience participation-based approach, New Delhi

Müller 1985, **Untersuchungen zur Sensibilisierung der lokalen Bevölkerung** für eine aktive Beteiligung an agroforstlichen Maßnahmen, Göttingen

Müller-Glodde 1994, **Der Runde Tisch als Programm?** Möglichkeiten und Grenzen der Institutionenförderung im Spannungsfeld von Umwelt und Entwicklung, Bonn: GTZ

Nohlen D, Nuscheler F 1993 (eds), Handbuch der Dritten Welt, Bd. 1, Bonn

O'Sullivan et al. **1979, Communication Methods to Promote Grassroots Participation**, Paris

OECD (early 2000), **Environmental Communication** - Applying Communication Tools towards Sustainable Development, Paris

OECD 1991, **Environment, Schools and Active Learning**, OECD Centre for Educational Research and Innovation, Paris

OECD 1993, **Environmental Education**. An Approach to Sustainable Development, ed. by Schneider, H. , Development Center Document, Paris

Oepen M (mid-2000), **MOVE – Moderation and Visualization for Participatory Group Events**. A Practical Guideline, Berlin

Oepen M 1986, **ACDC** - Appropriate Communication in Community Development, in: Communicatio Socialis Yearbook 5/1986, 87-102

Oepen M 1988, **Breaking the Culture of Silence**. A Development Support Communication Program in Indonesia, in: Sarilakas 3, Manila

Oepen M 1989, **Communicating with the grassroots**: a practice-oriented seminar series of DSE-ZEL, in: Group Media Journal 3/1990, S. 3-5

Oepen M 1990, **Community Communication**: The Missing Link between the 'Old' and the 'New Paradigm'?, in: Development Communication Journal, 1(1990)1, S. 52-63

Oepen M 1992, **Gold in the garbage**. Media support to Indonesian scavengers, in: development communication report, 76(1992)1, S. 16f

Oepen M 1992, **Gold in the Garbage**: Media Support to a Scavenger Development Program in Indonesia, in: D+C 4/92, S. 26f

Oepen M 1993, **Umweltbildung in Entwicklungsländern**. Gutachten zu Ansätzen bei UNESCO, OECD, WWF und CAP, Bonn: GTZ

Oepen M 1995 , **Community Communication**, in: Oepen (ed): Media Support and Development Communication in a World of Change, Bad Honnef, pp 40-54

Oepen M 1995 (ed), **Media Support and Development Communication in a World of Change**. Proceedings of a ACT/FU Berlin/ WIF conference in Berlin 1993, Bad Honnef

Oepen M 1996 , **Community Communication for Rural Development**. A DSE Training Program, Feldafing

Oepen M, Fuhrke U, Krüger T 1994, **Umweltbildung und Umweltkommunikation** - Erwartungen, Erfahrungen, Erkenntnisse, Wedemark

Padua S 1991, **Conservation awareness through an environmental education school program** at the Morro do Diabo State Park, São Paulo State, Brazil, Gainesville

Padua S 1995, **Environmental education programs for natural areas** in underdeveloped countries - a case study in the Brazilian Atlantic Forest, in: Palmer J et al (eds): Planning Education to Care for the Planet, pp 51-56

Padua S, Jacobson S 1993, **A comprehensive approach to an environmental education program in Brazil**, in: The Journal of Environmental Education 24/4, pp 29-36

Palmer J, Goldstein W, Curnow A 1995 (eds), **Planning education to care for the earth**, IUCN

Papen von U 2000, **Tse-Tse-Plage in Moroland**, in: Politische Ökologie 17(2000)63/64, p 29

Parker LJ 1997, **Environmental Communication**. Messages, Media & Methods. Dubuque, IA: Kendall/Hunt

Piotrow P et al 1997, **Health communication**: Lessons from family planning and reproductive health, Westport

Poffenberger M 1997, **Local knowledge in conservation**, in: IUCN, Beyond Fences, vol 2, Gland, pp 41-43

Poffenberger M, McGean B 1996, **Village Voices, Forest Choices**: Joint Forest Management in India, Oxford University Press, New Delhi.

Ramírez R 1997, **Cross-cultural communication and local media**, in: IUCN, Beyond Fences, vol 2, Gland, pp 111-114

Rice R, Atkin C 1989 (eds), Public Communication Campaigns, London

Richmond GM 1978, **Some Outcomes of an Environmental Knowledge and Attitude** Survey in England, in: Science and Education, vol 8

Rikken 1993, **Natural Resource Management by Self-Help Promotion** in the Philippines, Manila: GTZ,8/1993

Rikken 1993, **The Greening of Libertad**, Manila

Rohwedder WJ 1990 (ed), **Computer-Aided Environmental Education**. Troy, OH: The North American Association for Environmental Education

Salas et al. 1991, **Agricultural Knowledge Systems and the Role of Extension**, Hohenheim, 5/1991

Schade 1983, **Psychologische Aspekte umweltgerechter Entwicklungspolitik**, in: E+Z 10/83, S. 24f

Schneider H 1993 (ed), **Environmental Education**, Paris: OECD

Schönhuth M et al. 1993, **Partizipative Erhebungs-und Planungsmethoden in der Entwicklungszusammenarbeit**, Eschborn: GTZ

Schönhuth M, Kievelitz U 1994, **Participatory Learning Approaches**, GTZ

Scoones I, Thompson J 1994 (eds), **Beyond Farmer First: Rural People's Knowledge**, Agricultural Research and Extension Practice, Intermediate Technology Publications, London

Senge P 1994, **The Fifth Discipline Fieldbook**, New York

Serra 1990, **Community media in the Philippines**, in: group media journal, 3/1990, S. 10f

Shrestha AM 1987, **Conservation Communication in Nepal**, Kathmandu

Singhal A, Rogers EM 1999, **Entertainment-Education**: A communication strategy for social change, Mahwah

Sirait M 1995, **Mapping customary land**: a case study of Long Uli village East Kalimantan, Indonesia, as cited in: Poole P, Indigenous Peoples. Mapping & Biodiversity Conservation, Washington D.C., pp 20-24

Smith W 1995, **Behavior, social marketing and the environment**, in: Palmer J et al (eds): Planning education to care for the earth, Gland, pp 9-20

SPAN Consultants 1993, **Managing Education and Communication**. A manual for planners and advisors, Utrecht

Stuart 1989, **Access to media**: Placing video in the hands of people, in: Media Development 4/1989, S. 8-11

Sülzer 1980, **Medienstrategien und Entwick-lungspolitik**, in: Rundfunk und Fernsehen 28(1980), S. 56-69

Tehranian M 1997, **Communication, Participation and Development**, in: Servaes J et al (eds): Participatory Communication for Social Change, London, pp 44-63

Thioune 1993, **Environmental Education Techniques** Suitable for Use in Rural Areas in Senegal, Bradford, 8/1993

UNCED 1992, **Environment Screening Conference** for UNCED , New York, 5/ 1992

UNESCO 1990, **Environmental Education**: Selected Activites of UNESCO-UNEP Environmental Education Programme 1975-1990, UNESCO

UNESCO 1991, **Atelier de formation des techniciens et planificateurs** en charge de la conservation des sites naturels et pour les leaders des communautés paysannes. Projet pilote Mayombe, République du Congo

UNESCO 1992, **Environmental awareness building**, special issue 28(1992)2 of: Nature & Resources, a quarterly magazine by UNESCO

UNESCO 1992, **Gestion des ressources et des réserves de la bioshère et éducation relative à l'environment**. Projet pilote de Dja, Republique Cameroun

UNESCO (undated), **Environment and Development Information Kit**, contains: 5 issues of Environment and Development Briefs, 10 brochures on Environment and Development: Policy and Coordination, Capacity Building, Environmental Education and Information, Man and the Biosphere Programme, Biological Diversity, World Heritage, Fresh Water Resources, Oceanography, Coastal Marine Sciences, Earth Sciences

UNESCO-UNEP 1985, **A Problem-Solving Approach to Environmental Education**, UNESCO-UNEP Environmental Education Programme, Environmental Education Series 15

UNESCO-UNEP 1992, **Environmental Education Activites for primary Schools**. Suggestions for making and using low cost equipment, UNESCO-UNEP Envi. Education Programme, Envi. Education Series 21

UNESCO-UNEP 1992, **Report on Environmental Education Activities** 1/90-1/91 of the UNESCO-UNEP International Environmental Education Programme (IEEP)

UNESCO-UNEP 1993, **Publications of the UNESCO-UNEP International Environmental Education Programme (IEEP)**

Valladares-Padua C 1993, **The Ecology, Behavior and Conservation of the Black Lion Tamarin**, University of Florida

Weekes-Vagliani 1992, **Lessons from the Family Planning Experience** for Community-Based Env. Education, OECD Development centre, Technical Paper No. 62

Weltner K 1993, **Fachdidaktik der Physik**. Abschiedsvorlesung an der J.W. Goethe-Universität Frankfurt, unpubl. lecture, Oct 15

White 1989, **Networking and change in grassroot communication**, in: Group Media Journal, S. 17-21

Wibowo 1990, **Motivating Indonesian villagers through traditional theatre and flanoflex**, case study presented to a DSE seminar on 'Communication and rural development', Feldafing

Wickramasinghe A 1997, **Gender concerns in conservation**, in: IUCN, Beyond Fences, vol 2, Gland, pp 22-25

Wilson 1993, **The Sistren Theatre Collective in Jamaica**, in: Lewis (ed): Alternative Media: Linking Global and Local, Paris: UNESCO 1993, S. 41-50

Woerkum van C, Arts N 1998, **Communication between farmers and government about nature**: A new approach to policy development, in: Röling N.G., Wagemakers M.A.E: Facilitating Sustainable Agriculture, Cambridge

Wolff HP 1997, **Beurteilung der Zielgruppenbeteiligung in Entwicklungsvorhaben**, GTZ-MA-CILSS, Ouagadougou

Worldview 1991, **Worldview International Foundation Report** 1991-1992

WWF 1993, **Environmental Education Catalogue**

Young JE 1994, **Using Computers for the Environment**, in: L Starke (ed): State of the World 1994. New York, NY: W.W. Norton & Co., pp 99-116

Selected Case Studies

Formal Education

Madagaskar	WWF: Environmental Education as national programme through teaching material and teacher´s training for primary and secondary schools, not much broad effect (von Loebenstein 1993)
Sahel	EG+CINAM: Training and Information Programmes on the Environment for teachers /pupils/parents in nine west-african countries through curricula, teacher´s training, learning material, community action; not much broad effect (Filho 1993)
World	UNESCO: International Environmental Education Programme, offers regional specific training material, curricula and teacher´s training to interested member countries (Oepen 1993)

Non-formal Education

Burkina Faso	Patecore: positive example of the role of NGO and participation for the improvement of village land use (Funke 1992)
Ecuador	Fundacion Natura: Longterm lobby and sensitization measures in several steps for industrial and political decision makers in environmental and consumer´s protection; curricula development, teacher´s training, 150 radio programmes, visual media, community development (OECD 1993: 215)
Ghana	EPC+IEED: Environmental Education Strategy for Ghana, concluding formal and informal education, community development, NGO, media, advisory services, religious institutions and a catalogue of aims and acitivities. Governmental, little participation (Education o.J.)

India	Centre for Environment Education: formal and non-formal environmental education e.g. 'News and Feature-Service' for 1.000 periodica and a journalist network, curricula, teacher´s training, (OECD 1993: 53)
Kenya	Kengo (Kenya Energy and Environment Organisation): umbrella organization of 200 environmental NGOs with own publication, seminars, lobbying, training, community development; (OECD 1993: 48)
Kenya	Wildlife Clubs of Kenya: environment clubs at 1.500 (or 77% of all) secondary and primary schools with magazine, newsletter, publications, teacher´s training, similar to Chongololo/Zambia, Mila Hai Clubs/Tanzania, Wildlife Clubs/Uganda; (OECD 1993: 57)
Malaysia	CAP (Consumers Association Penang): indirect environmental education through seminars, courses, media, lobbying, legal assistance, consumer´s education and consulting, consumer´s clubs, teacher´s training, exposure for pupils, students, NGO, public authorities, enterprises etc. (Oepen 1993; Keiper 1985; CAP 1989)
Nepal	GTZ: Media Support for Waste Management via campaigns, horizontal communi-

	cation with citizens´s partici-pation; (Waste 1992)
Nepal	KMTNC: Formal und Infor-mal Environmental Commu-nication via home visits, semi-nars, visual media, adult edu-cation, curricula 6th-8th grade in Annapurna Conservation Area Project; (von Loebenstein 1993)
Peru	several NGO: Environmental Education and Awarenessbuilding via in-service teacher´s training and network (APECO). 'School, Ecology and the Peasant Com-munity' with radio-campaigns for teachers, pupils, daily 1 hour rural radio, Andean Rural School for peasant leaders, cadres; action oriented; (OECD 1993: 233)
Sahel	IUCN: Environmental Educa-tion and Communication, through Walia-Magazine (5.000 issues) from/for pupils/teachers (5.000/400) in order to make pupils nature protec-tors and reach adults via them. Redaction meeting twice per year with 25 schools and action oriented nature clubs; similar IUCN-approaches in Niger, Burkina Faso, Senegal; (OECD 1993:159)
Senegal	ENDA: Micro-project Ap-proach to Environmental Education, rural youth organi-zations and 12 primary schools with workshops result-ing in mini-projects on hy-giene, health, agro-forestry, stock-farming and on social aspects; (OECD 1993: 137)
Thailand	PDA: Environmental Educa-tion and youth work via seminars, visual media, video at schools and in the project center of Rural Development

	for Conservation Project; (von Loebenstein)
Thailand/Nepal	GTZ: Manual for Urban Environmental Management Nepal includes in the action plan training for administrative bodies, NGO, teachers, re-searchers, electronic and traditional media, NGO-networks; (Manual 1993)
World	NGO-co-operation with School Sector and Journalists as e.g. Fundacion Natura/Ecuador, Living Earth/Cameroon for formal educa-tion, CAP/Malaysia, YIH/Indonesia for non-formal education, Panos, Worldview International Foundation, Centre for Environment Edu-cation/India for media report-ing; (OECD 1993: 45; Worldview 1992)
World	School Magazine: Action/Southern Africa, Chongololo/WCSZ-Zambia, Piedcrow/CARE-Kenya, Walia/IUCN-Mali, Tortoise/Ghana etc; (OECD 1993)
Zimbabwe	Zimtrust: Campfire. Role of participation, NGO and legal title for resource management by rural population and district administration; (von Loebenstein 1993)

Awareness raising

Bolivien	SEMTA+GTZ: Comics and Picture Stories with high identification value and mobi-lization to bridge the intercul-tural gap between peasants and consultants; (gate 2/93, S. 12+ 22)
Cote d'Ivoire	WWF: In vain PR for Tai National Park in Buffer Zone and for tourists with slide-voice-show, video, film and T-shirts, to protect the park from

	environment destructing use; (WWF Environmental Education Dossier 1992)	**Communication**
Ecuador	**Fundacion Natura: Longterm lobby and sensitization measures in several steps** for industrial and political decision makers in environmental and consumer´s protection; curricula development, teacher´s training, 150 radio programmes, visual media, community development (OECD 1993: 215)	

Bolivien — **SEMTA+GTZ: Comics and Picture Stories** with high identification value and mobilization to bridge over intercultural barriers between peasants and consultants; (gate 2/93, S. 12+ 22)

Brasil — **WWF: Communication for Conservation**. In vain trial to protect a park from destructive use by using slide-voice-shows, movies and T-shirts (Dietz 1992)

Ecuador — **Fundacion Natura: Longterm lobby and sensitization measures in several steps** for industrial and political decision makers in environmental and consumer´s protection; curricula development, teacher´s training, 150 radio programmes, visual media, community development (OECD 1993: 215)

India — **Dasholi Gram Swarajya Mandal: Eco-development camps** 3-4 times p.a. for 250-300 concerned people, planers, govenment, fieldworker, NGO, academics, teachers in damaged mountain regions for mutual experience exchange and new action orientation; (OECD 1993: 59)

China — **'China Environment News': national daily newspaper** 3x/ week with 1/2 Mio no. of copies since 1984 and 400 associated journalists; (OECD 1993: 49)

Indonesien — **YHI: Environment Sensitization** at Environmental Centre Seloliman through sensual experience, seminars and training for different social groups; (WWF Environmental Education Dossier 1992)

Costa Rica — **Broadcasting and Education Ministry: Environment Education via Adult Education on Broadcast;** (Thioune 1993: 52)

Dominican Republic — **Broadcast: Environment Education via 'Interactive Radio'** for schools and audition clubs; (Thioune 1993: 53)

Togo — **CFSME: Environment Analysis via picture stories** with high identification value and mobilization potential; identical with GRAPP-method in Westafrica; (Thioune 1993: 67, Hoffmann 1991; Albrecht 1987: 2. Bd)

Gambia — **WIF: Sensitization of rural population for Environment Protection** via traditional media, video, horizontal communication; teacher/pupil as motivators; (Worldview 1992)

Zimbabwe — **IUCN: Zimbabwe Environmental Awareness Support Programme** supports NGO in conceptualizing and implementing of 'Environmental Awareness Camps' for teachers/pupil groups; additionally a pupil´s magazine with 100.000 issues; (unpub.)

India — **Centre for Environment Education: Formal and non-formal Environment Education**, e.g. 'News and Feature-Service' for 1.000 periodica und a journalist network, curricula, teacher´s training; (OECD 1993: 53)

India — **Kerala Sastra Sahitya Parishad/KSSP: Environmental Sensibilization and Lobby Work** through street theatre

	(250 times per month), village study circle, media campaigns, traditional scrolls etc. followed up by concrete local community development- und environmental care-measures, representation of interest around 'Silent Valley'; (OECD 1993: 49; Baskaran 1990)
India	**Chipko Movement with theater, traditional Media. Marshes** in Ghandi-Trad. (violent-free opposition) with 170 NGO over 400km with meetings, theatre, singing on resource protection in mountain region, followed up by concrete lokal community development- and environmental care-measures;(Baskaran 1990)
India	**Link Society** (NGO) with street theatre 1989 on a 1.000km-marsh through 300 villages and 8 cities on resource protection themes followed up by concrete local community development- and environmental care-measures; (OECD 1993: 61)
India	**SAC-DECU: 2000 traditional theatre programme for markets** to environment problems through urbanization; (Baskaran 1990)
Indonesia	**GTZ: Integrated Media Strategy** towards waste disposal/ recycling via theatre, video, TV under participation of affected people; (Oepen 1992)
Indonesia	**PPLH: ACDC - Traditional theatre** for proliferation of adopted land use methods for peasants within a participatory Area Development Programm; (Oepen 1986)
Kenya	**WIF: 'Social Marketing for 'Indigenous Food Plant Programme' and** biodiversity through traditional media,
	video, bulletins, schools und community development; (Worldview 1992)
Malawi	**GTZ: Traditional Theatre instead of Video for Health Education;** (Hollenbach 1993)
Malaysia	**WWF: Mobile Unit Conservation Programme**. in vain trial related to the National Park in Saba, to convince 50 villages within three years to adapt environmental sound landuse via slide-voice-shows, discussions and films; (OECD 1993: 60)
Nigeria	**National Conservation Foundation und TV Authority: Environmental Education via TV-doll theatre,** although as british import; (Doran 1994)
Pakistan	**IUCN: Journalist Resource Centre for the Environment** with 'News and Feature-Service' for journalist network, seminars, training, media campaign, prize competition, quiz, etc. (OECD 1993: 53)
Ruanda	**GRAPP: Picture Stories** with high identification value and mobilization potential also for participative environment analysis at and from rural population groups; similar to **CFSME/Togo** and Westafrica; (Hoffmann 1991; Gabathuler 1991)
Sri Lanka	**WIF: Environm. Education via TV Quiz,** school calendar and books, NGO-media training (Worldview 1992)
Southern Africa	**Action Magazine: Environment Magazine** for pupils/ teacher, 80.000 number of copies, comics, quiz etc, partly with follow-up via local media (WWF Environmental Education Dossier 1992)
Southern Africa	**IUCN: 'Communicating the Environment' via Mass Media**, Database, NGO network

287

	for decision makers, multiplikators (unpublished mission-report of 1994)
Thailand	**Thai Environment and Community Development Association: 'Magic Eyes'-Multimedia Campagne** 1987 about urban environment pollution and waste reduction, later in rural areas about forest destruction; similar to **Bangladesh Government on World Environment Day** 1989 about tree plantings, environment protection; (OECD 1993: 61)
Thailand	**WIF: Broad- and narrowcasting for Highland Development.** Environment protection through auditor´s clubs in Dialects, 2-way-communication between villages and administrative bodies, video, NGO-mediatraining (Worldview 1992)
World	**Journalist Associations: Environmentally engaged Journalist Networks**, normally mass media on national (e.g. Nepal/Zimbabwe/Zambia Association of Environmental Journalists), regional (Asia-Pacific Forum) and international level; (Oepen 1993)
World	**Panos Institute, London: With 'Down to Earth', 'Panoscope', 'Panos Feature Service'** a world-wide net on production, distribution und publication of environmental informations (unpublished Mission-Report of BMZ)
Zambia	**Wildlife Conservation Society Zambia: Environment Magazine and Broadcast** plus auditor´s clubs for pupils/teachers partly with follow-up via local media (unpublished. Mission-Report of BMZ)

Contributors

Institutional Partners

ACT - Appropriate Communication in Development
in Wedemark, Germany

Deutsche Gesellschaft für Technische Zusammenarbeit (GTZ) GmbH,
Pilot Program on Institutional Development in the Environment (PVI)
in Bonn, Germany

World Conservation Union - IUCN, Social Policy Group and
Commission on Education and Communication (CEC) in Gland, Switzerland

World Bank, Knowledge Products and Outreach Division
in Washington D.C., USA

Authors

Ronny Adhikarya, a Ph.D. in communication from Stanford University, USA, at present Senior Training Officer at the Knowledge Products and Outreach Division of the World Bank, in Washington D.C., USA. He served for 15 years at the United Nations Food and Agriculture Organization (FAO) as a senior-level specialist in charge of extension, education and training. He has worked with a wide network of training institutions and development specialists in 37 countries worldwide in the areas of sustainable agricultural development, environment, population, health, AIDS prevention, etc. He has written seven books on communication, extension and non-formal education subjects.

Noelle Arts is working as a research assistant at the University of Wageningen, Netherlands.

Karola Block worked for the Deutsche Gesellschaft für Technische Zusammenarbeit (GTZ) GmbH from 1996 to 1998 as an environmental adviser, planner, evaluator, author and trainer for the Pilot Program on Institutional Development in the Environment.

Marco A. Encalada is the general manager of Corporación OIKOS, Ecuador, a private consulting company, and a member of the Commission on Education and Communication of IUCN.

Wendy Goldstein, a graduate in Science and Education with a postgraduate qualification in environmental education from Sydney University and the College of Advanced Education in Australia, is currently Head of the Environmental Education and Communication Program of IUCN, the International Conservation Union which coordinates a program of an international network of communication and education experts: the Commission on Education and Communication (CEC). This program advocates the integration of communication and education as an instrument of policy and undertakes capacity-building for the education/communication staff in government and non-governmental organizations, particularly amongst IUCN members.

A Gomis is working as a research assistant at the University of Wageningen, Netherlands.

Winfried Hamacher holds an MSc in Electrical Engineering and Social Sciences from the Universities of Düsseldorf and Bochum, Germany. He worked as a teacher and as a free-lance lecturer from 1982 to 1987, and at a professional school in Ouagadougou, Burkina Faso from 1988 to 1990. He has been working for the Deutsche Gesellschaft für Technische Zusammenarbeit (GTZ) GmbH since 1991, currently acting as an environmental adviser, planner, evaluator, author and trainer for the Pilot Program on Institutional Development in the Environment.

Frits Hesselink started his career as a fellow at the Institute for International Law of the University of Utrecht, became involved in curriculum development for law and social studies, and taught various subjects at the State High Schools in Utrecht. He has been managing director since 1983 of SME MilieuAdviseurs, the Institute for Environmental Communication, and was involved for more than a decade in the formulation and implementation of various Dutch national programs for environmental education. Since 1994, he has chaired the IUCN Commission on Education and Communication and, since 1999, he has worked for his own consultancy company in the field of environmental education, communication and training, carrying out projects for government and international organizations all over the world.

Ulrich Juedes is working at the Institute of Science Education – IPN at the University of Kiel, Germany, currently specializing in sustainable development research.

A. Soedradjat Martaamidjaja is currently working at the Agency for Agricultural Education and Training (AAET) in Jakarta, Indonesia.

Gillian Martin Mehers is the Director of International Training at LEAD International (Leadership for Environment and Development), an international non-profit institution working to create a global network of leaders committed to the principles of sustainable development. Ms Martin Mehers has been involved in curriculum and materials development, training design and has worked as a facilitator and trainer, particularly in the areas of environmental communication planning, team-building and networking, and environmental conflict management.

Manfred Oepen completed communication and political studies at the Freie Universität Berlin, Germany and at Stanford University in the United States. Between 1984 and 1993, he managed projects with integrated media support systems in Indonesia in both rural and urban community development and environmental contexts. Since 1992, he has been general manager of a private consulting company, ACT - Appropriate Communication in Development. He has provided consultancy services for development and environmental communication-related projects, and developed and implemented related training curricula for German and international organizations in more than 20 countries all over the world.

Suzana M. Padua is the President of IPÊ - Instituto de Projetos e Pesquisas Ecológicas in Brazil, and a member of the Commission on Education and Communication of IUCN.

Saumya Pant is a doctoral student at the School of Interpersonal Communication, Ohio University. His research interests center around communication and national development, entertainment-education and social change, and ethnography.

Mariam Rikhana is currently working at the Agency for Agricultural Education and Training (AAET) in Jakarta, Indonesia

Everett M. Rogers, one of the founding fathers of development communication, has been a professor at Iowa State University, Stanford University and, currently, the University of New Mexico. His 1962 research on the "diffusion of innovation" is the second most cited work in the Social Sciences. He has written numerous publications on the role of communication in development processes, and shaped related policies as a consultant to UN organizations, the World Bank, and other organizations in around 50 countries all over the world. Recently, his teaching and research interests have centered on the entertainment-education communication strategy.

Arvind Singhal is an Associate Professor at the School of Interpersonal Communication, Ohio University. His teaching and research interests focus on the entertainment-education communication strategy, organizing for social change, and diffusion of innovations. He has served as a consultant to organizations such as the World Bank, UNFAO, UNAIDS, UNDP and others.

William Smith, Ed.D. in non-formal education from the University of Massachussetts, is the Executive Vice President and Division Director of Social Development at the Academy for Education in Development – AED in Washington D.C., USA.

Monique Trudel, a bioecologist by training, has fifteen years of international experience in environmental education and communication (EE&C). She worked as Regional Advisor for the IUCN EE&C Program to West Africa from 1985 to 1993. Currently, she is a free-lance consultant working on planning, training, monitoring and evaluation of public education in environmental programs on behalf of UN organizations, NGOs and IUCN.

Thomas Zschocke, instructional designer at Bridgewater State College, Massachusetts, USA, is completing his doctoral research on Web-based training at the Center for International Education, University of Massachusetts Amherst, USA.

Cees van Woerkum is working as a research assistant at the University of Wageningen, Netherlands.

Abbreviations

AAET	Agency for Agricultural Education and Training
ABC	Applied Behavioral Change
ACT	Appropriate Communication in Development
AED	Academy for Education in Development
AHSD	Agricultural High School for Development
AIR	All India Radio
AISTC	Agricultural In-service Training Center
AIT	Asian Institute of Technology
AV	Audiovisual
BIRC	Bandung Integrated Recycling Center
CAE	College for Agricultural Extension
CAMPFIRE	Communal Areas Wildlife Management
CBG	Community-based Group
CBO	Community-based Organization
CEC (IUCN)	Commission on Education and Communication
CEC	Continuing Education Center
CFSME	Conscientisation-Formation-Stimulation-Moyens-Evaluation (Visualization and extension method used in Africa)
CGI	Common Gateface Interface
CIAD	Centre for Integrated Agricultural Development
CM	Conflict management
CMC	Computer-mediated communication
COCE	Conference on Communication and Environment
CSC	Cascading Style Sheet
CSS	Community Self-survey
DAC	Development Assistance Committee
DAD	Decide, Announce and Defend
DAIEC	District Agricultural Information and Extension Center
DSE	Deutsche Stiftung für Internationale Entwicklung (German Foundation for International Development)

EC De Peel	Environmental Cooperative de Peel
ECEP	Environmental Communication and Education Plan
EE&C	Environmental Education and Communication
EET Netw.	Networking on Environmental Education and Training
EETC	Environmental Education, Training and Communication
EETM	Environmental Education and Communication Training Modules
EIA	environmental impact assessment
EnvCom	Environmental Communication
FAO	Food and Agriculture Organization
FEW	Field Extension Worker
FOKUS	Forum Komunikasi Daur Ulang Sampah (Communication Forum on Recycling)
GEF	Global Environment Facility
GRAPP	Groupe de Recherche et d'appui pour l'autopromotion Paysanne (Visualization and extension method used in Africa)
GreenCOM	USAID Environmental Education and Communication Project
GTZ	Deutsche Gesellschaft für Technische Zusammenarbeit (GTZ) GmbH
HTML	Hypertext markup language
IAE	International Academy for the Environment
IIRR	International Institute of Rural Reconstruction
IKIP	Bandung Teachers College
IPE	Instituto de Projetos e Pesquisas Ecológicas
IPM	Integrated Pest Management
IPN	Institute of Science Education (University of Kiel)
IRIS	Information Recall and Impact Survey
IUCN	International Union for Conservation of Nature and Natural Resources
KAP	Knowledge, Attitude and Practice
LEAD Int.	Leadership for Environment and Development

M&E	Monitoring and Evaluation	**TPS**	Tempat Persampahan Sementara (Waste ransfer station)
METAP	Mediterranean Environmental Technical Assistance Program	**TVE**	Television for the Environment
MOVE	Moderation and Visualization for Participatory Group Events	**UNAIDS**	Joint and Cosponsored United Nations Program on HIV/AIDS
NFEE	Non-formal environmental education	**UNCED**	United Nations Conference on Environment and Development
NGO	Non-governmental Organization	**UNDP**	United Nations Development
NPP	Natural Policy Plan		Program
OECD	Organization for Economic Cooperation and Development	**UNEP**	United Nations Environment Program
PA	Public Awareness	**UNESCO**	United Nations Educational,
PCD	Participatory curriculum development		Scientific and Cultural Organization
PDK	Perusahan Daerah Kebersihan (Municipal Sanitation Company in Bandung, Indonesia)	**UNFPA**	United Nations Fund for Population Activities
		UNICEF	United Nations Children's Fund
PEC	Primary environmental care	**URL**	Uniform Resource Locater
PEDEXTRA	Population Education through Agricultural Extension Training	**WBP**	Committee to preserve de Peel
		WHO	World Health Organization
PPP	Planning/Process/Product	**WPRDP**	Western Pichincha Regional Development Program
PRA	Participatory Rapid Appraisal		
PRSSNI	Network of private radio stations in Indonesia	**WWF**	World Wide Fund for Nature
		ZEL	Zentralstelle für Ernährung und Landwirtschaft (Center for Food and Agriculture of the German Foundation for International Development)
PUEA	Participatory Urban Environmental Appraisal		
PVI	Pilotvorhaben Institutionenentwicklung im Umweltbereich (Pilot Program on Institutional Development in the Environment of GTZ)	**ZÖV**	Zentralstelle für Öffentliche Verwaltung (Center for Public Administration of the German Foundation for International Development)
REA	Rapid Environmental Appraisal		
REC	Rural Extension Center		
RT/RW	Rumah Tangga/Rumah Warga (Lowest adminstrative unit in Indonesia)		
SD	Sustainable Development		
SDM	Studio Driya Media, an NGO in Indonesia		
SDRE	Extension, Education and Communication Service		
SEC	Strategic Extension Campaign		
SIA	Social Impact Assessment		
SMS	Subject Matter Specialist		
SWM	Solid Waste Management		
SWOT	Strengths, Weaknesses, Opportunities and Threats		
TOT	Training of Master Trainers		
TPA	Tempat Persampahan Terakhir (Final dump site)		

Umweltbildung, Umweltkommunikation und Nachhaltigkeit
Environmental Education, Communication and Sustainability

Herausgegeben von/Edited by Walter Leal Filho

Peter Lang · Europäischer Verlag der Wissenschaften

Thomas Immelmann / Otto G. Mayer /
Hans-Martin Niemeier / Wilhelm Pfähler (eds.)

Aviation versus Environment?

Frankfurt/M., Berlin, Bern, Bruxelles, New York, Wien, 2000. 143 pp.
ISBN 3-631-35874-1 · pb. DM 49.–*
US-ISBN 0-8204-4389-1

Aviation versus Environment? Late public discussion of this issue did not
even put a questionmark. The suspected vast effects of air traffic on global
environmental problems such a climate change or ozone layer depletion led
to an urgent demand for both practical solutions and a political framework
for the conduct of airports and airlines. This book contains the expertise of
environmental organisations, politicians, airline and airport managers who
discussed these issues on the second Hamburg Aviation Conference, held
on February 17 – 19, 1999. Taking into account the environmental and
economic needs for air traffic, the conference was a stage for a constructive
debate between aviation business and environmentalists and showed a
broad range of measures in regard of environmental problems linked with
aviation.

Contents: Global Air Transport and Ecological Limits · International and
European Environmental Policy: Protection of Industry or Environment? ·
Environmental Policy of Airports in the Age of Privatization

Frankfurt/M · Berlin · Bern · Bruxelles · New York · Oxford · Wien
Distribution: Verlag Peter Lang AG
Jupiterstr. 15, CH-3000 Bern 15
Fax (004131) 9402131
*incl. value added tax
Prices are subject to change without notice.